Lecture Notes in Mathematics

Edited by A. Dold and B. Eckmann

1263

V. L. Hansen (Ed.)

Differential Geometry

Proceedings of the Nordic Summer School
held in Lyngby, Denmark Jul. 29 – Aug. 9, 1985

Springer-Verlag
Berlin Heidelberg New York London Paris Tokyo

Editor

Vagn Lundsgaard Hansen
Mathematical Institute
The Technical University of Denmark
Building 303, DK-28000 Lyngby, Denmark

Mathematics Subject Classification (1980): 32C 10; 32H 10; 53C 05; 53C 20;
53C 55; 53C 80; 58E 05; 58E 11; 58E 20; 58G 30; 81E 10

ISBN 3-540-18012-5 Springer-Verlag Berlin Heidelberg New York
ISBN 0-387-18012-5 Springer-Verlag New York Berlin Heidelberg

This work is subject to copyright. All rights are reserved, whether the whole or part of the material is concerned, specifically the rights of translation, reprinting, re-use of illustrations, recitation, broadcasting, reproduction on microfilms or in other ways, and storage in data banks. Duplication of this publication or parts thereof is only permitted under the provisions of the German Copyright Law of September 9, 1965, in its version of June 24, 1985, and a copyright fee must always be paid. Violations fall under the prosecution act of the German Copyright Law.

© Springer-Verlag Berlin Heidelberg 1987
Printed in Germany

Printing and binding: Druckhaus Beltz, Hemsbach/Bergstr.
2146/3140-543210

Preface

During the period 29 July - 9 August 1985 a Nordic Summer School in mathematics was held at the Technical University of Denmark in Lyngby. The subject of the summer school was "Differential Geometry with Applications". The school was attended by 81 participants and had 8 main lecturers. In addition to the main lectures, there was a series of seminars by the participants.

The main reason for choosing differential geometry as the subject for the 1985 Nordic Summer School in mathematics was that the last two decades have witnessed a new strong interaction between mathematics and physics originating in the study of differential geometry of fibre bundles in mathematics and field theories in physics. For many physicists the goal is to find a theory combining all known forces of nature, and it appears that the so-called gauge theories, which in the language of differential geometry correspond to connections in fibre bundles, may provide an answer.

Any reasonable field theory is based on a variational principle, and the extremals for the associated differential geometric variational problem are the solutions searched for. Normally, this leads to problems concerning nonlinear (partial differential) equations on manifolds.

The purpose of the summer school was to present some of the important mathematical tools, methods, and results necessary for doing research within differential geometry and its above mentioned applications in theoretical physics.

The overall structure of the final program for the summer school contained 3 main themes (sections):

> **Gauge theory/moduli:** General variational methods and other global analytic methods to describe spaces of solutions to nonlinear problems from differential geometric physics.
> *Main lecturers:* Jean Pierre Bourguignon and Cliff Taubes.

> **Twistor methods:** Methods based on the theory of harmonic and holomorphic maps for constructing specific solutions to nonlinear equations arising from differential geometric variational problems.
> *Main lecturers:* Francis Burstall and John Rawnsley.

> **Global differential geometry:** This section contained some of the foundations for the applications in the two preceding ones, in particular elements of the theory of complex manifolds (especially Kähler manifolds). However, it also contained topics of independent interest,

e.g. comparison theorems in Riemannian geometry, and partial differential equations arising in differential geometry.

Main lecturers: Robert Greene, Karsten Grove and Jerry Kazdan.

This volume contains the manuscripts for the main lectures given at the summer school, except for the lectures by Cliff Taubes.

The opening address was delivered by Jim Eells, who spoke on the physics of classical field theories. This was further supplemented in a second lecture by Eells and a lecture by Peter Braam, who spoke on the quantum aspects. On our request, Peter Braam has kindly written a manuscript, which opens this volume and serves as an introduction to the section on gauge theory.

The program for the summer school was planned in close collaboration with Karsten Grove. Also Jim Eells, John Rawnsley and Jean Pierre Bourguignon were extremely helpful advisors. In connection with the practical arrangements, Flemming Damhus Pedersen was of immense importance. Splendid work was also done by our secretarial staff, not least Lone Aagesen, who was attached to the project from the outset. In the preparation of this volume, Steen Markvorsen has been of great help to me.

The summer school was made possible through donations from Nordiska Forskarkurser, the Danish Natural Science Research Council, Otto Mønsteds Fond and the Danish Mathematical Society.

Lyngby, March 1987   Vagn Lundsgaard Hansen

Table of Contents

Preface	III
List of participants	VII
List of seminars	XI

P. BRAAM
    Quantum field theory: the bridge between mathematics and the physical world........ 1

J.P. BOURGUIGNON
    Yang-Mills theory: the differential geometric side.......... 13

F. BURSTALL
    Twistor methods for harmonic maps............................ 55

J. RAWNSLEY
    Twistor methods.. 97

J.L. KAZDAN
    Partial differential equations in differential geometry...... 134

K. GROVE
    Metric differential geometry................................. 171

R. GREENE
    Complex differential geometry................................ 228

Differential Geometry with Applications
Nordic Summer School 29. July - 9. August 1985
The Technical University of Denmark, Lyngby

List of participants

Uwe Abresch, University of Bonn, Germany.
Lars Smedegaard Andersen, University of Århus, Denmark.
Amir Assadi, University of Virginia, USA.

Gunnar Berg, University of Uppsala, Sweden.
Göran Bergqvist, University of Linköping, Sweden.
Jonas Bjerg, Technical University of Denmark.
Per-Anders Boo, University of Umeå, Sweden.
Jean-Pierre Bourguignon, École Polytechnique, Palaiseau, France.
Peter Braam, University of Utrecht, Netherlands.
Francis Burstall, University of Bath, England.

Ulf Carlsson, University of Lund, Sweden.
Carlos M. Corona, University of Pennsylvania, USA.

Knut Dale, Telemark Distriktshøgskole, Norway.
Celso Melchiades Doria, University of Warwick, England.
J. Dupont, University of Århus, Denmark.

James Eells, University of Warwick, England.
Henrik Egnell, University of Uppsala, Sweden.
Thomas Erlandsson, University of Uppsala, Sweden.

Lisbeth Fajstrup, University of Århus, Denmark.
Niels Fjeldsø, University of Oslo, Norway.
Bent Filipsen, University of Århus, Denmark.
Juhani Fiskaali, University of Oulu, Finland.

Robert Greene, University of California at Los Angeles, USA.
Karsten Grove, University of Maryland, USA.
Thomas Gunnarsson, University of Luleå, Sweden.
Lars Gæde, Technical University of Denmark.

Heikki Haahti, University of Oulu, Finland.
Johan Hamberg, University of Lund, Sweden.
Dominique Hulin, École Polytechnique, Palaiseau, France.

Yoe Itokawa, Max-Planck-Institut für Mathematik, Bonn, Germany.

Kurt Johansson, University of Uppsala, Sweden.

Henrik Karstoft, University of Århus, Denmark.
Devendra A. Kapadia, University of Oxford, England.
Jerry L. Kazdan, University of Pennsylvania, USA.
Ryoichi Kobayashi, Max-Planck-Institut für Mathematik, Bonn, Germany.
Torbjørn Kolsrud, University of Stockholm, Sweden.
P.B. Kronheimer, University of Oxford, England.
Martti Kumpulainen, University of Oulu, Finland.
M. Kuukasjarvi, University of Oulu, Finland.
Osmo Kurola, University of Oulu, Finland.

Hans Lindblad, University of Lund, Sweden.
Sauli Luukkonen, University of Oulu, Finland.

Ib Madsen, University of Århus, Denmark.
Christophe Margerin, École Polytechnique, Palaiseau, France.
Steen Markvorsen, Technical University of Denmark.
Gregor Masbaum, University of Nantes, France.
Nina Morishige, University of Oxford, England.
Jesper Michael Møller, University of Copenhagen, Denmark.
Eljas Määttä, University of Oulu, Finland.

Sæmundur Kjartan Ottarsson, University of Warwick, England.
Ulf Ottoson, University of Gothenburg, Sweden.
Hijazi Oussama, Max-Planck- Institut für Mathematik, Bonn, Germany.

Erik Kjær Pedersen, University of Odense, Denmark.
Henrik Pedersen, University of Odense, Denmark.
Jan Pedersen, University of Århus, Denmark.
Osmo Pekonen, University of Jyväskylä, Finland.
Peter Petersen, University of Maryland, USA.
Peter Petterson, University of Lund, Sweden.
Yat Sun Poon, University of Oxford, England.

Hans-Bert Rademacher, University of Bonn, Germany.
Hans Rang, University of Stochholm, Sweden.
John Rawnsley, University of Warwick, England.
Jaime Brinck Ripoll, IMPA, Rio de Janeiro, Brasil.
Ryszard Rubinsztein, University of Uppsala, Sweden.
B. Michael Rumpf, Technical University of Denmark.
Torbjørn Ryeng, University of Linköping, Sweden.

Alexander P. Stone, University of New Mexico, USA.
Henrik Svensmark, Technical University of Denmark.

Cliff Taubes, University of California, USA.
Klaus Thomsen, University of Arhus, Denmark.
Gudlaugur Thorbergsson, University of Bonn, Germany.
Anne Marie Torpe, University of Odense, Denmark.
Tor Tunheim, Alta Lærerhøgskole, Norway.

Giogio Valli, University of Pisa, Italy.
Marina Ville, École Polytechnique, Palaiseau, France.

Anatol Winter, Technical University of Denmark.
Carl Aage D. Winther, Technical University of Denmark.
Christopher M. Wood, University of Southampton, England.
John C. Wood, University of Leeds, England.

Bent Ørsted, University of Odense, Denmark.

List of seminars

U. Abresch	H-surfaces and explicit solutions of the sinh-Gordon equation.
U. Abresch	Lower curvature bounds, Toponogov's theorem and bounded topology.
A. Assadi	Some remarks on curvature, symmetry and differentiable structure.
J. Dupont	Dilogarithms as characteristic classes.
J. Gravesen	Spaces of holomorphic mappings and instantons.
H. Haahti	On linear connections given by projection valued maps.
O. Hijazi	A conformal lower bound for the smaller eigenvalue of the Dirac operator and Killing spinors.
Y. Itokawa	Around the sphere theorem - a survey.
R. Kobayashi	Moduli space of generalized K3 surfaces with a Kähler-Moišezon metric.
P.B. Kronheimer	Magnetic monopoles and multi-Taub-NUT metrics.
S. Markvorsen	Heat kernel comparison on minimal submanifolds.
E. Määttä	On connection preserving maps between vector bundles over Banach and Hilbert manifolds.
H. Pedersen	Einstein metrics, spinning top motion and monopoles.
Y.S. Poon	Self-duality on $P^2 \# P^2$.
A.P. Stone	Remarks on electromagnetic lenses.
G. Thorbergsson	Tight immersions of 4-dimensional manifolds.
M. Ville	Pinching the sectional curvature in dimension 4.
J.C. Wood	The construction of harmonic maps from surfaces to Grassmannians.
C.M. Wood	Harmonic sections.

Quantum field theory: the bridge between

mathematics and the physical world

by

Peter Braam

§ 0 Introduction.

These notes, based on a talk on the subject, should serve one purpose only: make mathematicians enthusiastic for quantum field theory. Thus they have been written in a heuristic rather than a completely rigorous style, in fact many topics mentioned in § 2 have yet to be rigorously formulated.

We have stressed the central role that symplectic geometry plays in the theory, and the analogy between ordinary quantum mechanics and the quantization of field theories. Therefore, in section 1 classical mechanics is rather quickly repeated, and in section 2 quantization is treated. We would like to thank Dr. P. v. Baal for long discussions on the subject.

§ 1 Mechanics and momentum maps.

The importance of mechanics lies not only in its physical origin, but also in the fact that mechanical ideas are widespread in physical quantum theories and in mathematics. To describe a mechanical system one can choose a certain setting, each with its own advantages. Here the following two are used:

1) Lagrangian formulation

Motion is a path, $c(t)$, in a configuration manifold M, and the path which is actually followed by our mechanical system is a critical point of the action functional, which is a map from the space of paths to $\mathbb{R}$, defined by:

$$S(c) = \int L(c(t), \dot{c}(t)) dt \qquad 1\text{-}1$$

where $L: TM \to \mathbb{R}$ is the Lagrangean. Now L is not an arbitrary element of $C^\infty(TM)$ and we require it to be of the form

$$L(m, X_m) = g_m(X_m, X_m) - V(m) \qquad 1\text{-}2$$

where $g(\cdot,\cdot)$ is a Riemannian metric on M, called <u>kinetic energy</u>, and $V: M \to \mathbb{R}$ is the <u>potential</u>. The critical paths $c(t)$ are solutions of the <u>Euler Lagrange</u> equations:

$$\frac{\partial}{\partial t}\left\{\frac{\partial L}{\partial \dot{q}_j}(c(t),\dot{c}(t))\right\} = \frac{\partial L}{\partial q_j}(c(t),\dot{c}(t)) \quad j=1,\ldots,n \qquad 1\text{-}3.$$

Here $(q_1,\ldots,q_n)$ are local coordinates around $c(t)$ on M.

The physical meaning of the action functional may seem mysterious at this stage, but quantum mechanics sheds some light on this, see §2, and Manin [9], Ch. 3.

Before we proceed to give the Hamiltonian or symplectic formulation, we discuss an example:

<u>Celestial mechanics.</u> Consider n particles in $\mathbb{R}^3$ with masses m_i. The configuration space is $M = \mathbb{R}^{3n}$ and let q_{ki} be coordinates on M with $k=1,2,3$, $i=1,\ldots,n$. Then:

$$\begin{aligned}L(q,\dot{q}) &= \sum_i \frac{1}{2} m_i \|\dot{q}_i\|^2 + \sum_{i<j} G m_i m_j \|q_i - q_j\|^{-1} \\ &= \sum_{k,i} \frac{1}{2} m_i \dot{q}_{ki}^2 + \sum_{i<j} G m_i m_j \left(\sum_{k=1}^{3}(q_{ki}-q_{kj})^2\right)^{-1/2}\end{aligned} \qquad 1\text{-}4$$

($G > 0$, the gravitational constant). The first term on the rhs of (1-4) clearly represents the kinetic energy, whereas the second is the gravitational potential. Most facts in mechanics were discovered solely through consideration of the equations of celestial mechanics:

$$\frac{d}{dt} m_i \dot{q}_{ki} = m_i \ddot{q}_{ki} = -\sum_{i<j} \frac{(q_{ki}-q_{kj}) G m_i m_j}{\left(\sum_\ell (q_{\ell i}-q_{\ell j})^2\right)^{3/2}}, \quad i=1,\ldots,n,$$

the Euler-Lagrange equations of (1-4).

2) Hamiltonian formulation

The advantage of Hamiltons version is that the place (q_i) and speed $(\dot{q}_j)$ variables no longer play a different role. Put differently the theory is mathematically a bit more intrinsic.

Let X be a <u>symplectic manifold</u>, i.e. a manifold X with a so-called <u>symplectic form</u>, $\omega \in \Omega^2(X)$, where ω obeys

a) $\omega_x : T_xX \times T_xX \to \mathbb{R}$ is nondegenerate

b) $d\omega = 0$.

A mechanical system is now described by a <u>Hamiltonian</u> or <u>total energy function</u> $H \in C^\infty(X)$. This Hamiltonian H determines a vectorfield V_H on X, called the <u>Hamilton vectorfield</u> of H, defined as follows:

$$i_{V_H} \omega = dH \ . \quad (i_. \text{ is contraction}) \qquad 1\text{-}5$$

<u>Hamiltons equations</u> now read

$$\dot{x}(t) = V_H(x(t)) \ . \qquad 1\text{-}6$$

If we put $X = T^*M$ then we can recover the Euler Lagrange equations. For convenience we work in coordinates q_i on M. The conjugate momenta:

$$p_j = \frac{\partial L}{\partial \dot{q}_j} = \frac{\partial g(\dot{q},\dot{q})}{\partial \dot{q}_j}$$

supply coordinates on $T^*M = X$ and it is not hard to see that

$$\omega = \sum_j dp_j \wedge dq_j$$

defines (globally!) a symplectic form on X. Associated to a Lagrangian $L(q,\dot{q})$ is the Hamiltonian

$$H(q,p) = \sum_j p_j \dot{q}_j(p) - L(q, \dot{q}(p))$$

Now the Hamilton vector field is given by:

$$V_H(q,p) = \left(+\frac{\partial H}{\partial p} \ , \ -\frac{\partial H}{\partial q} \right)$$

so Hamiltons equations read:

$$\dot{q} = \frac{\partial H}{\partial p}$$
$$\dot{p} = -\frac{\partial H}{\partial q} \ . \qquad 1\text{-}7$$

The equivalence of 1-7 and 1-3 follows easily from the definitions of p and H. A more extended exposition of Hamilton's and Lagrange's description of mechanics can be found in Abraham-Marsden [1] and Arnold [3].

Finally we discuss conserved quantities in mechanical systems in the setting of momentum maps, see also Bourguignon's seminar notes. Let G be a connected Lie group acting symplectically on (X,ω) ; this means $g^*\omega = \omega$ ($g \in G$) . Consider the vectorfields induced by the action, i.e. consider the map ($\mathfrak{g}$ is the Lie algebra og G)

$$\tilde{} : \mathfrak{g} \to \Gamma(TX): Y \to \tilde{Y} = \left\{ m \to \frac{d}{dt}(\exp tYm)\big|_{t=0} \right\}$$

Then for $Y \in \mathfrak{g}$, $L_{\tilde{Y}}\omega = d(i_{\tilde{Y}}\omega) = 0$, so assuming for simplicity that $H^1(X, \mathbb{R}) = 0$, we get:

$$i_{\tilde{Y}}\omega = d\Phi_Y \quad \text{for some} \quad \Phi_Y \in C^\infty(X)$$

in other words $\tilde{Y}$ is the Hamilton vectorfield of Φ_Y . Now Φ_Y is only determined up to a constant, but with a bit more work we get a map:

$$\Phi : X \to \mathfrak{g}^* \quad \text{s.t.} \quad <\Phi, Y> = \Phi_Y \quad , Y \in \mathfrak{g}$$

and: 1-8

$$\Phi_Y(g \cdot x) = \Phi_{Ad_{g^{-1}}Y}(x) \quad g \in G$$

(Φ is almost unique, see Abraham-Marsden [1] Ch. 4). Φ , first defined by Lie, is called the <u>momentum map</u> of the G action, and Φ determines the complete G action by integration of the Hamilton vectorfields $\tilde{Y} = V_{\Phi_Y}$.

From 1-8 we see that the co-adjoint orbit of $\Phi(x) \in \mathfrak{g}^*$ is constant on the orbit of $x \in X$ (because $\Phi(gx) = {}^t Ad_{g^{-1}} \Phi(x)$). If G is Abelian we get that $\Phi(x)$ is constant on Gx . In general $\Phi^{-1}\{0\}$ is a G invariant set and

$$\Phi^{-1}\{0\} / G \qquad \qquad 1-9$$

is called the <u>symplectic quotient</u> or <u>Marsden-Weinstein</u> reduction of $X: \Phi^{-1}\{0\} / G$ is symplectic itself.

<u>Example</u>: Given a Hamiltonian H on (X,ω) whose flow is complete. Here $\mathbb{R}$ acts symplectically by:

$$(t,x) \to \phi(t,x) \in X$$

where $\phi(t,)$ is the flow of V_H . Clearly for $1 \in \mathfrak{g} = \mathbb{R}$ we have $\tilde{1} = V_H$ so

$$\Phi : X \to \mathbb{R}^* : x \to H(x)$$

is constant on the $\mathbb{R}$ orbits. Here we recovered <u>conservation of energy</u>.

§2 Classical field theories and their quantization.

In §1 classical mechanics was presented. Here our interest lies in the quantization of this, leading to quantum mechanics. Next we introduce classical field theories, more specifically gauge theories. These can be quantized just as classical mechanics, but the original, classical, Hamiltonian system is now infinite dimensional. The theory has not yet found a mathematically mature framework; our considerations will be heuristic.

To quantize classical mechanics we give a scheme of geometric quantization (see Abraham-Marsden [1] 5.4). The arrows ↔ mean something like "corresponds to".

	Classical mechanics		Quantum mechanics
(i)	Points $(q,p) \in X = T^*\mathbb{R}^k$	↔	Functions ϕ in Hilbert space $L^2(\mathbb{R}^k)$, called <u>wave functions</u>.
	More precisely in symplectic terms:		
	Points in a symplectic manifold (X, ω)	↔	Sections of a $\mathbb{C}$-line bundle L with connection ∇, which are cov. constant on the leaves of a Lagrangean foliation of X.

This process is called <u>geometric quantization</u>.

(ii)	Certain functions on X	↔	Self adjoint operators on $L^2(\mathbb{R}^k)$ called <u>observables</u>.
	Example:		
	place functions q_j	↔	$\phi(x) \to x_j \phi(x)$
	momentum " p_j	↔	$\phi(x) \to i\hbar \frac{\partial \phi}{\partial x_j}$

Here $\hbar$ is a small positive number called <u>Planck's constant</u>. The limit $\hbar \to 0$ should describe classical mechanics.

(iii) paths $q(t), p(t)$ ⟷ A function $\phi(x,t)$ on $\mathbb{R} \times \mathbb{R}^k$
s.t. $\phi(-,t) \in L^2(\mathbb{R}^k)$.

initial values $p(0), q(0)$ ⟷ A function $\phi(-,t) \in L^2(\mathbb{R}^k)$

Hamiltons equations in a ⟷ a) <u>Schrödingers equation</u>:
potential $V(x)$ $$i\hbar \frac{\partial \phi}{\partial t} = \hbar^2 m \Delta \phi + V(x)\phi$$

b) The <u>propagator</u> or <u>Greens function</u> K defined by:
$$\phi(x,t) = \int_{\mathbb{R}^k} K(x,t;y,0)\,\phi(y,0)\,dy$$

a) and b) are equivalent: see below.

So far our scheme. The idea of quantum mechanics is that physics is described better using the quantum theory than the classical theory. The wave function $\phi(x,t)$ is supposed to describe the system, not really the states of the system; these aren't so directly accessible. Information of a probabilistic kind about the states can be obtained through the observables, e.g.:

(i) Expectation values of coord. q_j at time t is:
$$\int_{\mathbb{R}^k} \phi(x,t)\, x_j\, \overline{\phi(x,t)}\, dx = <\phi, Q_j \phi>_{L^2}.$$

(ii) Heisenberg's uncertainty relation: (a consequence of the set up)

$$\begin{pmatrix}\text{variance in observed}\\ \text{values of } q_j\end{pmatrix} \cdot \begin{pmatrix}\text{variance in observed}\\ \text{values of } p_j\end{pmatrix} \geq \hbar.$$

Note that in particular (ii) shows a drastic difference with classical physics. Further details of quantum mechanics can be found in Dirac's book [5].

Before we turn to classical field theories we discuss the propagator, following Lee [8] ch. 19. The propagator is of course the heart of quantum mechanics because it describes the dynamics.

First write Schrödinger's equation in the form:
$$\frac{\partial \phi}{\partial t} = \frac{1}{\hbar i} H(\phi)\,,\quad H(\phi) = \hbar^2 m \Delta \phi + V(x)\phi.$$

It follows that

$$\phi(-,t) = \exp(t H / \hbar i) \phi(-,0) ,$$

so $K(-, t; -, 0)$ is just the integral kernel of $\exp(t H / \hbar i)$. Feynman, following a suggestion of Dirac, has given an explicit formula for $K(x, t; y, 0)$ in terms of a <u>path integral</u>:

$$K(x, t; y, 0) = \int_{\substack{\text{paths } q(s) \\ q(0)=y, \ q(t)=x}} \exp\left(i S(q(s))/\hbar\right) dq(s) \qquad 2\text{-}1$$

In this formula $S(q(s))$ is the classical action of the path $q(s)$ and $dq(s)$ is a so-called Gauss-Wiener measure on the space of paths. (Wiener had derived a similar formula for the heat equation earlier).

Integrals if this kind have been defined rigorously, see Simon [11], but we take a heuristic view at them, using a finite dimensional approximation of the space of paths. Pick an $N \in \mathbb{N}$, put $k = 1$ for simplicity, and approximate a path $q(s)$ by the "step-path" (fig. 2.1):

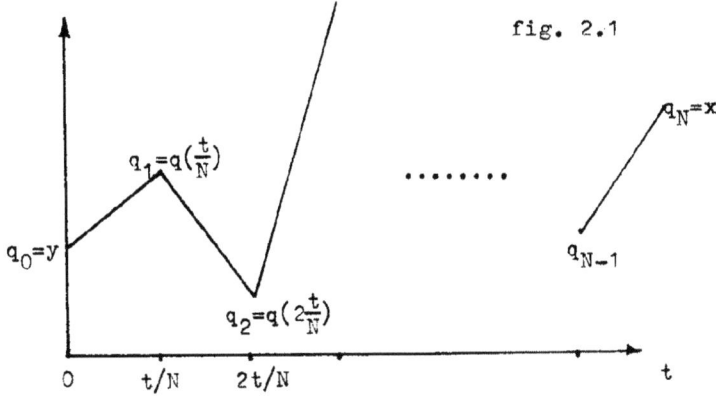

fig. 2.1

Now (2-1) can be computed as:

$$\lim_{N \to \infty} C_N \cdot \int_{\mathbb{R}^{N-1}} \exp\left[\frac{i}{\hbar} \frac{t}{N} \sum_{j=1}^{N} \left\{\frac{1}{2} m \left(\frac{q_j - q_{j-1}}{t/N}\right)^2 - V\left(\frac{q_j + q_{j-1}}{2}\right)\right\}\right] dq_1 \cdots dq_{N-1}$$

$$2\text{-}2$$

with $C_N = (2\pi i \hbar t/(N-1)m)^{-(N-1)/2}$. A fancy proof that $K(x,t;y,0)$ is indeed the kernel of the Schrödinger propagator can also be given along the line of a step path argument. Remark that in (2-2) the integrals are never absolutely convergent! According to some physicists, the advantage of using the Feynman integrals instead of the Schrödinger equation is, that they contain only the action, which is usually (but not here) a Lorentz invariant object.

Next we introduce <u>gauge theories</u> in vacuum. Let $\mathbb{R}^{3,1}$ be Minkowski space, G a compact Lie group and P a principal G-fibre bundle over $\mathbb{R}^{3,1}$. The configuration space here is not a space of paths but the space $C(P)$ of <u>connections</u> or <u>gauge potentials</u> on P. The action YM: $C(P) \to \mathbb{R}$ is called the <u>Yang-Mills functional</u> and is defined by

$$YM(A) = \int_{\mathbb{R}^{3,1}} \|F^A\|_M \, dV_M \qquad 2\text{-}3$$

where the subscript M denotes "Minkowskian", and

$$F^A = dA + \frac{1}{2}[A,A] \qquad 2\text{-}4$$

is the <u>curvature</u> or <u>field strength</u> of A. The corresponding Euler Lagrange equations are the so called <u>Yang-Mills equations</u>:

$$d^A *_M F^A = 0 . \quad (*_M \text{ is Minkowski Hodge star}). \qquad 2\text{-}5$$

The Yang-Mills functional is invariant under the action of an infinite dimensional Lie group $GA(P)$ on $C(P)$. This group is called the <u>gauge group</u> and defined by:

$$GA(P) = \Gamma(P \times_{Ad} G) .$$

A <u>gauge transformation</u> $g \in GA(P)$ sends $A \in C(P)$ to

$$g \cdot A = L_{g*} dg^{-1} + Ad_g A . \qquad 2\text{-}6$$

This implies $F^{gA} = Ad_g F^A$ and explains invariance of the action.

<u>Example:</u> Maxwell's electromagnetism in vacuum. Here $G = U(1)$ so A and F are up to a factor i just ordinary forms. As $F^A = dA$ we get

$$d\, F^A = d\, dA = 0 \qquad 2\text{-}7$$

$$d * F^A = d * dA = 0 . \qquad 2\text{-}8$$

Writing $dA = E_i \, dx^i \wedge dt + B_{ij} \, dx^i \wedge dx^j$ it is easy to see that up to physical constants 2-7 and 2-8 are the four Maxwell equations in vacuum. The action of $g = e^{i\phi} \in GA(P)$ on A is given by

$$g(iA) = g\,dg^{-1} + g\,i\,A\,g^{-1} = i(A - d\phi)$$

and this provides the principle that adding "a gradient" doesn't affect electromagnetism.

To achieve quantization of a gauge theory, i.e. to get a <u>quantum-field-theory</u>, we first switch to a Hamiltonian picture. For simplicity we restrict ourselves to electromagnetism, but the discussion can readily be generalized, see Arms [2] and Itzykson Zuber [7] 9-3.

Let C be the space of gauge potentials on $\mathbb{R}^3$, so $C = \Omega^1(\mathbb{R}^3)$. Then $T^*C \simeq \Omega^1(\mathbb{R}^3) \times \Omega^2(\mathbb{R}^3)$ via the integral pairing. For a curve $A(t)$ in C we define an action

$$S(A(t)) = \int L(A(t), \dot{A}(t))\,dt \qquad 2\text{-}9$$

with

$$L(A, \dot{A}) = \frac{1}{2}\|\dot{A}\|^2_{L^2,\mathbb{R}^3} - \frac{1}{2}\|dA\|^2_{L^2,\mathbb{R}^3} \qquad 2\text{-}10$$

If we let the path $A(t)$ in C correspond to the $\mathbb{R}^{3,1}$ gauge potential $\tilde{A}(x,t) = \pi^*A(t)$, with $\pi: \mathbb{R}^{3,1} \to \mathbb{R}^3$ the space projection then 2-9, 2-10 is precisely the Yang-Mills action 2-3, $YM(\tilde{A})$. The <u>electric</u> and <u>magnetic field</u> can be found as: $E = \dot{A}(t)$, $H = dA$. The conjugate momentum of $\dot{A}(t) = E$ is the so-called electric displacement:

$$D = *E = *\dot{A}(t) \quad (*\text{ the } \mathbb{R}^3 \text{ Hodge star})$$

so this gives rise to a Hamiltonian H:

$$H: T^*C \to \mathbb{R}: (A,D) \to H(A,D) = \frac{1}{2}\|D\|^2_{L^2} + \frac{1}{2}\|dA\|^2_{L^2}.$$

The corresponding Hamilton equations are:

$$\dot{A} = *D \qquad 2\text{-}9$$

$$\dot{D} = d*dA = d*H = dB = \text{rot } B \qquad 2\text{-}10$$

Now 2-9 results in $d\dot{A} = d*D = \text{rot } E$, a Maxwell equation; 2-10 is another and a third can be found by noting that $dd A = dH = \text{div } H = 0$, but $\text{div } D = 0$ is missing.

To find this missing sheep, we note that $GA = C^\infty(\mathbb{R}^3, U(1))$ acts in a Hamiltonian way on T^*C as follows:

$$(e^{i\phi}, (A,D)) \to (A - d\phi, D) \in T^*C .$$

The momentum map $\Phi : T^*C \to ga^* \simeq \Omega^3(\mathbb{R}^3)$ is given by

$$(A,D) \to dD = \text{div } D \in \Omega^3(\mathbb{R}^3) \simeq ga^* .$$

So we should work on $\Phi^{-1}\{0\}$ to get the fourth Maxwell equation div D = 0 satisfied. However, to quantize we need a symplectic manifold and this can be achieved rather naturally now by going over to the <u>Marsden Weinstein reduction</u> $C_{phys} = \Phi^{-1}\{0\}/GA_o$ (GA_o component of 1 of GA)*. So what we see is that the <u>constraint $\Phi = 0$</u> on our Hamilton equations, <u>naturally brings</u> with it that only <u>gauge invariant objects</u> should be considered in doing quantum theory. This is an important observation in building up a quantum field theory, see Itzykson Zuber [7] 9-3.

We may now take a look at the Feynmann integrals for a gauge theory. From 2-1 we see that they must look like:

$$\int_{\left\{\begin{array}{l}\text{paths in } \Phi^{-1}\{0\}/GA\\ \text{with boundary cond.}\end{array}\right\}} \exp\left\{\frac{i}{g} S([A](t))\right\} d[A](t) . \qquad 2\text{-}11$$

here g is a constant > 0 , called the <u>coupling constant</u>. Now an important result, the so-called Ward identities, that 2-11 can be written as:

$$\int_{\left\{\begin{array}{l}\text{connections on } [0,t] \times \mathbb{R}^3\\ \text{with boundary cond.}\end{array}\right\}/GA_{\dim 4}} \exp\left\{\frac{i}{g} YM(A)\right\} d[A] \qquad 2\text{-}12$$

i.e. as an integral over connections on a <u>four</u> manifold modulo the gauge group in dimension four. The derivation of 2-12 from 2-11 involves some calculations using δ-functions of the set $\{\Phi = 0\}$, see L. Faddeev [6]. The <u>Euclidean postulate</u> now states that 2-12 must be computed by replacing t by $-i\tau$. From 2-2 and 2-3 one easily deduces that this amounts to replacing the Minkowski "norm" of F^A by the Euclidean one, and here is the birth of the <u>Euclidean gauge theories</u>. For the Euclidean postulate see Itzykson, Zuber [7] 6-2, there it is referred to as <u>Wick rotation</u>.

---

* We have omitted the o from GA_o in the sequel for simplicity.

Boundary conditions in $\mathbb{R}^3$ and a loose approximation lead to compactifying $\mathbb{R}^3 \times [0,t]$ to a compact Riemannian four manifold M on which one considers slightly more general objects than 2-12 namely:

$$W(\chi) = \int_{C(P)/GA(P)} \chi(A) \exp\left\{+\frac{1}{g} S(A)\right\} d[A] \qquad 2\text{-}13$$

where $P \to M$ is a, usually nontrivial, G bundle and $\chi : C(P) \to \mathbb{C}$ is a GA invariant function. YM is now given by:

$$YM(A) = \int_M \|F^A\|^2 \, dV \quad . \qquad 2\text{-}14$$

Indeed, most physically relevant quantities can be expressed in the form 2-13. It is easy to see the physical relevance of <u>moduli spaces of instantons</u>. Recall that these moduli spaces are exactly the set of minima for YM on $C(P)/GA(P)$. Now for small g, 2-13 will get its main contribution from a neighbourhood of this moduli space. In fact in a rough approximation 2-13 reduces to an integral over this moduli space, see Schwarz [10].

Finally we would like to comment on some aspects of computing 2-13. Usually the integration is not carried out on $C(P)/GA(P)$ but on a slice S, for the $GA(P)$-action on $C(P)$, within $C(P)$. The slices used in physics, intersect $GA(P)$ orbits many times: the <u>Gribov ambiguity</u>. This is related to the topology of $C(P)/GA(P)$. In making this change of coordinates from $C(P)/GA(P)$ to S one encounters a Jacobian-determinant, however not a finite dimensional one. This must be <u>regularized</u>, and removing these "<u>infinites</u>", consistent with scaling properties, is the process of <u>renormalization.</u>

In computations with chiral fermions it happens that $\chi(A)$ is a determinant of a Dirac operator. To get a consistent quantum theory, a $GA(P)$ invariant regularization must exist, and this is not always possible. Phenomena of this kind are the <u>anomalies</u> and these again relate to the topology of $C(P)/GA(P)$. References for the topics mentioned here are: Singer [12], Itzykson Zuber [7] and Atiyah Singer [4] respectively.

<u>Acknowledgement:</u> The author is on leave from the Mathematical Institute, Utrecht, The Netherlands. A British Council FCO scholarship, under which this work was partly carried out, is gratefully acknowledged.

§ 3 References.

[1] Abraham R., Marsden J.; <u>Foundations of Mechanics</u>, 2nd ed., Benjamin 1978.

[2] Arms, J.; The structure of the solution set for the Yang-Mills equations, Math. Proc. Comb. Phil. Soc. 90 (1981) p. 361.

[3] Arnold, V.I.; <u>Mathematical Methods of Classical Mechanics</u>, GTM 60. Springer 1978.

[4] Atiyah M.F., Singer I.M.; Dirac Operators Coupled to vector potentials, Proc. Natl. Acad. Sci. USA 81 (1984) 2597-2600.

[5] Dirac P.A.M.; <u>The principles of quantum mechanics</u>, 4th ed. Oxford University Press, 1982.

[6] Faddeev L.; in proc. of the Summer School les Houches, 1975.

[7] Itzykson C., Zuber Y-B.; <u>Quantum Field Theory</u>, Mc Graw Hill, 1980.

[8] Lee T.D.; <u>Particle Physics and an Introduction to Field Theory</u>. Hardwood Ac. Publishers, 1982.

[9] Manin Yu.I.; <u>Mathematics and Physics</u>, Progress in Physics 3, Birkhäuser, 1981.

[10] Schwartz A.S.; Instantons and Fermions, Comm. Math. Phys. 64 (1979)

[11] Simon, B.; <u>Functional Integration and Quantum Physics</u>, Academic Press, 1979.

[12] Singer I.M.; Some remarks on the Gribov Ambiguity, Comm. Math. Phys. 60 (1978) p.7.

Yang-Mills Theory:

The differential geometric side

by

Jean Pierre Bourguignon

PREFACE

In recent years, Yang-Mills theory which is the subject of this series of lectures has attracted great attention from both mathematicians and physicists, hence the introductory Lecture I on the outgrowth of gauge physics. Its basic set-up that we present in Lecture II has now been recognized as that of connections on bundles over space-time, the curvature being the field strength, hence is geometric in nature. The non-linearity of the field equations, the Euler-Lagrange equation of the Yang-Mills functional, is therefore familiar to geometers. The 4-dimensional theory is by far the most interesting because of the occurence of the notion of self-duality. We shall develop it in Lecture III.

The second variation of the Yang-Mills functional is the subject of Lecture IV. It has an interesting algebraic structure, which leads in dimension 4 to important geometric consequences about stable Yang-Mills connections, such as local minima of the functional.

This very briefly summarizes the organization of this part of the lectures at this Nordic Summer School. (For more details, see Table of Contents.)

General references on differential geometric aspects of Yang-Mills theory are:

M.F. ATIYAH, Geometry of Yang-Mills fields, Lez. Fermi, Ac. Naz. dei Lincei, Sc. Norm. Sup. Pisa (1979).

M.F. ATIYAH, N. HITCHIN, I.M. SINGER, Self-duality in 4-dimensional Riemannian geometry, Proc. Roy. Soc. London A, 362 (1978), 425-461.

D.D. BLEECKER, Gauge theory and variational principles, Global Analysis: pure and applied series, Addison-Wesley Pub. Co., Reading (1981).

J.P. BOURGUIGNON, H.B. LAWSON Jr., Stability and isolation phenomena for Yang-Mills fields, Comun. Math. Phys. 79 (1981), 189-230.

J.P. BOURGUIGNON, H.B. LAWSON Jr., Yang-Mills theory: its physical origin and differential geometric aspects, in Seminar on Differential Geometry ed. by S.T. Yau, Ann. Math. Studies n° 102, Princeton University Press, Princeton (1982).

T. EGUCHI, P.B. GILKEY, A.J. HANSON, Gravitation, gauge theories and differential geometry, Phys. Rep. 66 (1980).

D.S. FREED, K.K. UHLENBECK, Instantons and four-manifolds, Math. Sci. Inst. Publications 1 (1984), Springer.

C.H. GU, On classical Yang-Mills fields, Phys. Rep. 80 (1981).

In the analytical approach, which is compulsory as soon as one deals with non homogeneous space-times, some of the difficulties are typical of geometric problems. To make the system of Yang-Mills equations amenable to standard techniques, one is forced to break the invariance of the theory under an infinite dimensional group, the gauge group. In dimension 4, the theory is conformally invariant, hence the fact that, in the weak formulation of the theory, one falls in the limiting case of Sobolev inequalities. By now, very important basic results have been obtained from this point of view, Mainly by K. Uhlenbeck, and by C. Taubes. Of great importance to physicists have been moduli spaces of solutions to the Yang-Mills equations. Results on them are by now fairly complete. These results will be presented by A. Bahri[1] and C. Taubes[2] in their lectures.

Of no less importance are the far-reaching consequences within mathematics of Yang-Mills theory, especially in the realm of 4-dimensional differential topology. These achievements follow mainly from the work of S. Donaldson, and, combined with another outstanding work by M. Freedman, have led to the proof of the existence of infinitely many distinct differentiable structures on $\mathbb{R}^4$. It is believed that this story only begins!

Editorial note:
1) The lectures of Bahri were given by Bourguignon and Taubes.
2) Manuscripts for the lectures by Bahri and Taubes are not included in this volume.

TABLE OF CONTENTS

Lecture I. A brief overview of gauge physics

    1. Steps toward geometrization of physics: Maxwell's laws of electromagnetism and general relativity   1

    2. XXth century physics: a rich harvest of new phenomena and a quest for unification   2

    3. A Yang-Mills formulation of electromagnetism as introduction   5

    4. Some further comments on Yang-Mills theory   7

    5. The emergence of supersymmetric theories   8

    6. Some (epistemological?) conclusions   9

    References   11

Lecture II. Yang Mills equations

    1. The basic set-up of Yang-Mills theory   14

    2. The first variation formula   20

    3. The gauge group   22

    4. Yang-Mills equations as an elliptic system in the curvature   23

Lecture III. Special features of 4-dimensional Yang-Mills theory

    1. Self-duality   25

    2. An algebraic introduction to spinors in dimension 4   28

    3. A glance at characteristic classes   30

    4. Basic examples of self-dual connections over S^4 32

    References   33

Lecture IV. The second variation and stable Yang-Mills fields

    1. The second variation formula   35

    2. Special variations: the enlarged gauge group   37

    3. The basic stability theorem   39

    4. The self-duality theorems   40

    References   42

Lecture I. A BRIEF OVERVIEW OF GAUGE PHYSICS

In this lecture we present (very briefly) some of the ideas that led the physicists to consider gauge theories, and in particular Yang-Mills theories, as viable theoretical models.

1. Steps toward geometrization of physics: Maxwell's laws of electromagnetism and general relativity.

By unifying electricity and light, Maxwell made a great step towards giving unity to physical phenomena by showing that force fields and oscillatory phenomena can have a common origin. In a sense, this can be viewed as the highest point of classical physics.

From a mathematical point of view, Maxwell's equations for an electromagnetic field are rather subtle in that the algebraic objects hidden in them were quite sophisticated for that time. They appeared at a time where Grassmann developed the exterior algebra calculus. These efforts culminated in the concept of an electromagnetic 2-form in a 4-dimensional space-time (hence a 6-component object) which incorporates the electric field (a 3-component vector), and the magnetic field (a 3-component pseudo-vector). This became clear as Lorentz introduced the group of transformations preserving a metric of signature (+++-) on the vector space $\mathbb{R}^4$. Time was ready for relativity theory.

Notice, that in this new geometric setting (invariant under the Lorentz group), Maxwell's equations of the vacuum take a beautifully condensed form, namely, the electromagnetic field Ω is a harmonic 2-form on space-time, i.e., we have

$$d\Omega = 0 , \quad \delta\Omega = 0 .$$

A dissymmetry exists between the two parts of the system, the form being closed being reflected in the absence og magnetic sources.

General relativity is the next step which was taken (cf. [E-G]). In general relativity, describing the distribution of matter in space-time, a metric is thought to be a potential. The field equations relate a second order nonlinear expression in the potential, the Einstein tensor, with the stress-energy tensor giving the physical action. The Einstein tensor has a strong geometric structure. On the space of potentials, it is the gradient of a naturally defined functional, the total scalar curvature, i.e., the integral over space-time of the scalar curvature of the Lorentzian metric.

2. XXth century physics: a rich harvest of new phenomena and a quest for unification.

As it is well known, in the twentieth century, physics underwent some fundamental revolutions.

First of all, two physical interactions besides electromagnetism and gravity were discovered, the strong and weak interactions. The strong interaction is responsible for the cohesion of the nucleus, and the weak one for the β-decay. Both of them act only at very short distances.

To resolve the duality between waves and particles, quantum mechanics was introduced. This new way of thinking created great turmoil in the scientific community. Indeed, it was not easy to think of light as being at the same time an oscillatory phenomenon, and a flow of particles. Reality forced physicists to get precise rules to deal with these problems.

At this point, the world as described by physics appeared quite complex. Four separate models were necessary to take care of all physical interactions, and, depending on the scale at which one was thinking, one had to use classical or quantum rules. In this correspondance, particles are characterized by their wave functions as fields.

Many attempts to get a (more) unified picture of the world were made by A. Einstein, H. Weyl (cf. [W]), (and also E. Cartan) among others. They tried and used the most sophisticated (geometric) tools they could think of.

Much earlier (1918), inspired by the coordinate invariance of relativity theory, H. Weyl introduced the notion of gauge invariance of physical laws. In an attempt to unify electromagnetism and general relativity, he proposed to take a conformal metric as potential (cf. [Mr]). The conformal factor was thought of as a local factor which could be fixed by choosing a local reference system. Some time later (1929), he showed that it was the phase of the wave function representing the particle submitted to electromagnetic interaction which should be gauged. The key feature of gauge theories is again the invariance of the physics under a group, but in this case an infinite-dimensional one. To make this more precise we return to the classical electromagnetic field Ω mentioned above. Notice that since $d\Omega = 0$, we may express Ω as

$$\Omega = d\alpha$$

where α is a 1-form on $\mathbb{R}^4$ called the electromagnetic potential. The form α is defined only up to an exact form, i.e., we may replace α by $\alpha + df$ where f is any smooth function on $\mathbb{R}^4$. Such a replacement is called a change of gauge or a gauge transformation. The insensitivity of the physics to the group of gauge transformations lies at the heart of the matters. It is called the Principle of local invariance.

Another idea is suggested by this approach, namely adding an internal parameter space (on which the phase group U_1 acts) to the space-time variables. This idea was further developed in the thirties in the so-called Kaluza-Klein theories. (For modern account, one can consult [Ps].)

At about the same time, the importance of the intrisic momentum, the spin, of some particles such as the electron or the proton was recognized by Pauli. The purely geometric notion of spinors was earlier studied on mathematical grounds by E. Cartan. This discovery forced physicists to consider the wave functions of these particles to be spinor fields. The use of spinor fields and of the basic differential operator acting on them, the Dirac operator, has recently aroused great interest among mathematicians to study the internal geometry of manifolds (cf. for example [G-L]).

It is only much later that gauge theories (i.e., theories admitting a similar local invariance) were considered. In [M-Y], C.N. Yang and R.L. Mills proposed a gauge theory with symmetry group SU_2 as a classical model for strong interactions. They also explain how such a classical variational theory can help understanding the quantum effects. (For a more recent account on this, one can consult [Jw].) Indeed, strong and weak interactions have to be dealt with following the rules of quantum mechanics. This new approach looked very promising for renormalization (i.e., getting rid of infinities in evaluating quantum perturbations).

Although the original paper contained all the ingredients of modern non-abelian gauge theories, the success was far from immediate. Generalizations to more general invariance groups were quickly considered. The main trouble came from the conformal invariance of the whole theory which forced the Yang-Mills fields to be massless. It is only after Higgs showed how by a spontaneous breaking of the symmetry one could split off the field into a lower dimensional field and a massive object that the theory regained interest. Shortly after, Weinberg and Salam used a U_2- Yang-Mills theory to unify the electromagnetic and weak interactions. The exchange particles for weak interactions, the so-called intermediate bosons $(W^+, W^-,$ and $Z)$, which are analogous to the photons for electromagnetism, are massive and by now have been "observed".

Notice that Yang-Mills-Higgs theory is also relevant to other physical theories such as superconductivity (cf. [J-T] Chapter 1 for example).

It is only in the early seventies that it was recognized that the whole setting of Yang-Mills theory is that of connections over G-bundles over space-time where G is a Lie group taken to be the symmetry group of the interaction under study. The field strength could be identified with the curvature of the connection and the action with the L^2-norm of the curvature. Over topologically non trivial space-times, the theory of characteristic classes is also pertinent to describe the charge of the particle in the Yang-Mills sense. (A parallel between the concepts elaborated independently by mathematicians and physicists can be found in [Wu-Y].)

This spectacular coincidence brings techniques of global differential geometry at the heart of the formalism of modern theoretical physics.

3. A Yang-Mills formulation of electromagnetism as introduction.

The Yang-Mills formulation above can be considered a strict analogue of electromagnetism theory as follows. For electromagnetic theory, the relevant symmetry group is U_1. The potential $i\alpha$ with values in $i\mathbb{R}$, the Lie algebra of U_1, can be considered as a connection on a trivialized principal U_1-bundle over space-time. A gauge transformation is then a smooth map $\tau : \mathbb{R}^4 \longrightarrow U_1$ which can be written as $\tau(x) = \exp(-if(x))$. The transformed connection is $\alpha^\tau = \alpha + df$ (as above), and the associated field $\Omega = d\alpha$ is just the curvature of the connection of the line bundle. One sees right away that the field is always closed, and the field equations are obtained by writing down the Euler-Lagrange equation for the total energy $L(\alpha) = \frac{1}{2} \int \|d\alpha\|^2$. If $\alpha_t = \alpha + t\beta$ is a family of potentials, we note that

$$\frac{d}{dt} L(\alpha_t)\big|_{t=0} = \int (d\alpha, d\beta)$$

$$= \int (\Omega, d\beta)$$

$$= \int (\delta\Omega, \beta) \quad .$$

Thus, we have obtained Maxwell's equation for the field Ω.

It is natural to ask whether there is a physical interpretation of the connection, i.e., of the gauge potential. For many years it was thought that the electromagnetic potential was merely a mathematical artefact, convenient but physically meaningless. Then, in 1959 an experiment suggested by Y. Aharonov and D. Bohm, and performed for the first time by Chambers, revealed that in the absence of the field, the electromagnetic potential does play a role. In this experiment, one reflects a coherent beam of electrons in a closed path encircling a solenoid. This solenoid is considered as an infinite, perfectly insulated tube. Although the field outside the tube is zero, the phase shift caused by the self-interaction of the beam is observed to vary with the intensity of the current in the tube. This phase shift is simply interpreted as the holonomy transformation of the flat bundle, generated by parallel translation around the closed path. This interference phenomenon has a quantum origin.

Topological effects connected with characteristic classes have also their counterpart in electromagnetic theory. In 1930, P.A.M. Dirac introduced the notion of a magnetic monopole, an electromagnetic field with an isolated singularity in space. He observed that the integral of the field over a sphere surrounding the singularity in ordinary space (properly normalized) could take on non-zero integer values. These integers, of course, come from the first Chern class of the underlying U_1-bundle, and their non-vanishing proves the non-triviality of this bundle. The existence of magnetic monopoles still remains conjectural. Nevertheless, the value of the bundle formalism in electromagnetic theory is evident.

4. Some further comments on Yang-Mills theory.

In fact, in recent years, an $(U_1 \times SU_2 \times SU_3)$-Yang-Mills theory, the so-called standard model, has been extensively studied. It was supposed to provide the basic framework for unifying electromagnetic, weak and strong interactions. In a sense, this says that all these interactions have a geometric origin. Because of their importance in evaluating the quantum fluctuations, critical points of the Yang-Mills action have been looked for with great interest. Surprisingly, on the standard sphere S^4, viewed as the Euclidean compactification of Minkowski space, a complete description of the minimal solutions was provided by algebraic geometric methods thanks to the Penrose twistor transform. Namely, when translated into twistor language, these solutions become holomorphic objects. (For Penrose, this is part of a very large programme, cf. [P]). They describe the geometric complication of the vacuum which can be viewed as pseudo-particle configurations, the instantons, which remain confined (it is expected that this is a nonlinear effect connected with the nonabelian character of the symmetry group).

It is indeed an amazing fact that the mathematical results needed to obtain this description had just been proved (cf. [A1] for details). One of the most challenging problems which remains unsolved is to decide whether other critical points exist. To study critical fields over more general spacetimes, these methods are irrelevant. One must see analytical techniques which are made delicate by the conformal invariance of the set-up, and the presence of an infinite dimensional invariance group, the gauge group, a typical feature of gauge theories. Big progress has been made recently using (and developing) global analysis techniques. Ideas (and pressure) from physicists were decisive at some crucial points.

These results will of course be presented at this school in the lectures given by Bahri and C. Taubes.

It was a big event when S. Donaldson turned the whole game around. He used the space of minimal solutions of the Yang-Mills action on an unknown manifold to study a longstanding problem in differential topology. The crucial property of this 5-dimensional moduli space is that it realizes in a geometric way a cobordism between the manifold under study and a connected sum of complex projective spaces. The process by which the manifold is attached as one component of the boundary is precisely the one by which a pseudo-particle localizes over a point. (This will be the content of the last lecture by A.Bahri).

5. The emergence of supersymmetric theories.

Other symmetries play an important role in today's understanding of the elementary particles. They associate particles having different spins and sensitive to different types of interactions. The leptons such as the electron, or the muon, react only to electromagnetic and weak interactions whereas the quarks are also subject to strong ones. These theories are called supersymmetric and open the way to a grand unification in which one expects the symmetry group to contain that of the standard model. (For the moment being, the groups SU_5, SO_{10}, and E_8 are the favorites among physicists.) So far, there is no real experimental evidence of the relevance of these theories. One of their physical consequences of great importance would be the existence of magnetic monopoles. They would not be elementary particles (or fields) but rather special configurations of other elementary fields. Their expected masses are enormous. This would explain why they have not yet been observed.

Some specific models of this type have been studied. Among them special attention has been devoted to non-linear σ-models which correspond to the study of harmonic mappings with values in symmetric spaces, one of the major topics of the series of lectures by J. Rawnsley and F. Burstall at this school. A link with Yang-Mills models is expected.

From a mathematical point of view, supersymmetry is a new type of geometric structure, some people say of course a supergeometry. Instead of working on a manifold as in ordinary differential geometry, one directly works on its bundle of exterior forms, the Grassmann bundle. New viewpoints on deep mathematical questions are expected to come out of this. (One can consult [A2], [Wn], [H], or [Gr] for applications to Morse theory and the index theorem for example.)

6. Some (epistemological?) conclusions.

By the preceding developments, we hope to have convinced the reader that physics and geometry have a long and rich history in common.

The present situation appears to us as characterized by the following features. First of all, the concepts which form the framework of a large part of contemporary fundamental physics are more than ever the building blocks of differential geometry, even when they were developed on purely mathematical grounds. In the opposite direction, we are taught that physically pertinent objects must be looked into by mathematicians. Secondly, the shift of interest among differential geometers towards global problems finds its counterpart in physics. Moreover, in recent years, problems raised by physicists have required the use of some of the most advanced theorems in mathematics. Rarely in history, this phenomenon has been witnessed with such a strength.

Nevertheless, the aims of mathematicians and physicists remain different. Whereas mathematicians try to understand structures for themselves, physicists are after saying something pertinent about the real world. A typical example of this difference in attitudes is given by the following story (cf. [Ls] for details) which we feel is not out of place in this introductory lecture. While studying in the late thirties 4-dimensional Lagrangians which are quadratic in the curvature of a metric, C. Lanczos observed that one was no good to generate field equations since its Euler-Lagrange equation while varying the metric was identically satisfied. From a mathematical point of view, this means that the value of the action one derived from it on a compact manifold is an invariant of the differentiable manifold itself. Lanczos had just discovered the formula expressing the Euler characteristic in dimension 4 as the integral of a quadratic polynomial in the curvature, but did not make any comment to that effect. Although the paper containing these considerations appeared in the Annals of Mathematics, it remained unnoticed by mathematicians. A little later, S.S. Chern got the whole story of expressing characteristic numbers as curvature integrals, the famous Chern-Bonnet-Gauss theorems. This discovery was a decisive step towards establishing global Riemannian geometry as an important body of mathematics. Taking an expression of the physicist Dyson (cf. [D]), is that a "missed opportunity" or should we believe in the "unreasonable effectiveness of mathematics in the natural sciences" as E.P. Wigner puts it (cf. [Wr])?

(First hand comments on the matter of this lecture can be found in the Nobel lectures by S.L. Glashow (cf. [Gw], A. Salam (cf. [S]), and S. Weinberg (cf. [Wg]). Other possible references are [I] and [J].)

References.

[A1] M.F. ATIYAH, Geometry of Yang-Mills fields, Lez. Fermi, Ac. Naz. dei incei, Sc. Norm. Sup. Pisa (1979).

[A2] M.F. ATIYAH, Circular symmetry and stationary-phase approximation, in Colloque en l'honneur de Laurent Schwartz, Volume 1, Astérisque 131 (1985), 43-60.

[D] F. DYSON, Missed opportunities, Bull. Amer. Math. Soc. 78 (1972), 635-652.

[E-G] A. EINSTEIN, M. GROSSMANN, Entwurf einer allgemeinerten Relativitäts-theorie und einer Theorie der Gravitation, Z. für Math. und Phys. 62 (1913), I. Physikalischer Teil, 225-244, II. Mathematischer Teil, 244-261.

[G-L] M. GROMOV, H.B. LAWSON Jr., Spin and scalar curvature in the presence of a fundamental group I, Ann. Math. 111 (1980), 209-230.

[Gr] E. GETZLER, Pseudodifferential operators on supermanifolds and the Atiyah-Singer index theorem, Commun. Math. Phys. 92 (1983), 163-178.

[Gw] S.L. GLASHOW, Towards a unified theory: threads in a tapestry, Rev. Mod. Phys. 92 (1980), 539-543.

[H] G. HENNIART, Les inégalités de Morse (d'après E. Witten), Séminaire Bourbaki 83-84, Exposé 617, Astérisque.

[I] J. ILIOPOULOS, Unified theories of elementary particle interactions, Contemp. Phys. 21 (1980), 159-183.

[J] A. JAFFE, Introduction to gauge theories, Proc. Int. Cong. Helsinki (1978), 905-916.

[J-T] A. JAFFE, C. TAUBES, Vortices and monopoles, Progress in Phys. n°2, Birkhäuser, Boston (1980).

[Jw] R. JACKIW, Quantum meaning of classical field theory, Rev. Mod. Phys. 49 (1977), 681-706.

[Ls] C. LANCZOS, A remarkable property of the Riemann-Christoffel tensor in four dimensions, Ann. Math. 39 (1938), 842-850.

[M-Y] R.L. MILLS, C.N. YANG, Conservation of isotopic spin and isotopic gauge invariance, Phys. Rev. 96 (1954), 191-195.

[Mr] M.E. MAYER, D.D. Bleecker's book review, Bull. Amer. Math. Soc. 9 (1983), 83-92.

[P] R. PENROSE, The twistor programme, Rep. Math. Phys. 12 (1977), 65-76.

[Ps] R.S. PALAIS, The symmetric criticality principle, Proc. Hefei-Shanghai Symp. Differential Geometry and Differential Equations, (1981).

[S] A. SALAM, Gauge unification of fundamental forces, Rev. Mod. Phys. 92 (1980), 525-536.

[W] H. WEYL, Selecta, Birkhäuser, Basel (1955).

[Wg] S. WEINBERG, Conceptual foundations of the unified theory of weak and electromagnetic interactions, Rev. Mod. Phys. 92 (1980), 515-524.

[Wn] E. WITTEN, Supersymmetry and Morse theory, J. Differential Geom. 17 (1982), 661-692.

[Wr] E.P. WIGNER, The unreasonable effectiveness of mathematics in the natural sciences, Commun. Math. Phys. 13 (1960).

[Wu-Y] T.T. WU, C.N. YANG, Concept of nonintegrable phase factors and global formulation of gauge fields, Phys. Rev. D 12 (1975), 3845-3857.

Lecture II. YANG-MILLS EQUATIONS

1. The basic set-up of Yang-Mills theory.

The framework in which the theory develops is that of bundles over manifolds with structure group a compact Lie group G.

As it is well known, bundles can be approached in two ways, via principal bundles, or via associated bundles. We shall use the second one, merely as a matter of taste although for physical applications they correspond to quite different things.

For a physicist, the base space of the bundle represents space-time, the structure group the symmetry group of the interaction under consideration (see Lecture I). A principal G-bundle P describes the laboratory boundary conditions. Since a specific elementary particle is associated with a linear irreducible representation, say μ, the associated bundle $E = P \times_\mu F$ will be attached to a type of particles, its sections being precisely the wave functions of particles of this type in this experiment. The basic idea is to incorporate appropriate internal degrees of freedom for each interaction. As we mentioned earlier, this idea was first applied with success by H. Weyl in electromagnetism.

Let us give some specific examples of such situations. Let us begin with the general linear group $\mathbb{R}Gl_n$ which is the structure group of the principal bundle of linear frames on any manifold M of dimension n. The tangent bundle, and more generally tensor or exterior bundles, are associated with this principal bundle.

Besides U_1 that we already met, the simplest group we will consider is the group SU_2 of unitary matrices with determinant 1 whose Lie algebra is spanned by the Pauli matrices

$$\sigma_1 = \begin{pmatrix} i & 0 \\ 0 & -i \end{pmatrix}, \quad \sigma_2 = \begin{pmatrix} 0 & 1 \\ -1 & 0 \end{pmatrix}, \quad \sigma_3 = \begin{pmatrix} 0 & i \\ i & 0 \end{pmatrix}.$$

The group SU_2 can be viewed as the group of unit quaternions, hence has the topological type of a 3-dimensional sphere S^3. It appears naturally as the structure group of the Hopf fibration of a sphere of dimension $4q+3$ (sitting as a round sphere in $\mathbb{H}^{q+1}$) over the quaternionic projective space $\mathbb{HP}^q$. These fibrations are straightforward generalizations of the Hopf maps from odd spheres to complex projective spaces.

Two bundles naturally associated with a principal bundle P are of great importance, namely, $GP = P \times_{ad} G$ its automorphism bundle, and $gP = P \times_{Ad} g$ its infinitesimal automorphism bundle. (Here, g denotes the Lie algebra of G .) Notice that, if G is abelian, these bundles are trivial. Since we decided to work in the associated bundle formalism, we will mainly meet their images into the endomorphism bundle $E^* \otimes E$ that we denote respectively by GE and gE . If the representation μ is (resp. infinitesimally) faithful, GE (resp. gE) admits the Lie group G (resp. the Lie algebra g) as fibre.

In Yang-Mills theory, the basic objects are G-connections on the G-bundle E . We say that a connection ∇ is a G-connection if it preserves any further structure defined on E by G , e.g., a fibre metric if the representation μ is orthogonal, or a complex structure if μ is complex, etc..

Since we adopted the vector bundle formalism, from the many equivalent definitions of a connection, we use the following one. For us, a connection ∇ is a linear first order differential operator from the vector space $\Omega^0(M,E)$ of sections of E to the vector space $\Omega^1(M,E)$ of E-valued differential 1-forms satisfying the following axiom. For any function f on M and any section s of E ,

$$\nabla(fs) = f \nabla s + df \otimes s .$$

This condition amounts to saying that the principal symbol of ∇ which is a linear bundle map between $T^*M \otimes E$ and $T^*M \otimes E$ is the identity, hence the universal role played by connections in the theory of bundles. Generalizing the ordinary derivative of functions in many variables, connections are adequate tools to differentiate sections of a bundle.

A connection ∇ has a natural extension to any tensor bundle built on E , e.g., its dual bundle E^* , or its endomorphism bundle $E^* \otimes E$, etc.. This is very useful, and allows us to express the G-invariance property of the connection ∇ . Indeed, any section of a tensor bundle built on E which is defined by the G-structure is parallel with respect to ∇ , in other words is left invariant by the parallel transport defined by ∇. In the case of the fundamental representation of an SU_2-bundle for example, both a complex structure, a hermitian product and a complex volume element are parallel with respect to an SU_2-connection.

It easily follows from the definition that the difference between two G-connections is an element of $\Omega^1(M, \mathfrak{g}E)$. Hence, the space CE of G-connections is an affine space modelled after $\Omega^1(M, \mathfrak{g}E)$.

There is a useful tool which is naturally associated with a connection ∇ on a vector bundle E, namely the exterior differential d^∇ for E-valued differential forms. Indeed, if, for an E-valued differential k-form α which is decomposable (i.e., $\alpha = \beta \otimes s$ for an ordinary differential k-form β and a section s of the bundle E), one sets

$$d^\nabla \alpha = d\beta \otimes s + (-1)^k \beta \wedge \nabla s \quad ,$$

one gets a meaningful definition. But d^∇ does not turn $\Omega^*(M,E)$ into a complex. In this context, the curvature R^∇ of the connection ∇ is just $d^\nabla \circ d^\nabla$.

In physics, connections are called gauge potentials, and the curvature R^∇ is referred to as the field strength of the potential ∇. This is consistent with what we saw in Lecture I for electromagnetism.

One can easily express the dependence of the curvature upon the connection ∇, namely

(1) $$R^{\nabla+A} = R^\nabla + d^\nabla A + [A \wedge A]$$

where $[A \wedge A]$ is the image under the $\mathfrak{g}$-bracket map of $A \wedge A$ considered as a 2-form with values in $\mathfrak{g}E \otimes \mathfrak{g}E$.

<u>Proof of Formula (1)</u>: Recall that by our definition of the curvature for a section s of the bundle E, $R^\nabla(s) = d^\nabla(d^\nabla s)$.

Therefore, since for an E-valued differential k-form α,

$$d^{\nabla+A} \alpha = d^\nabla \alpha + (A \wedge \alpha) \quad ,$$

where $(A \wedge \alpha)$ is the E-valued (k+1)-form obtained by composing the $\mathfrak{g}E \otimes E$-valued (k+1)-form $A \wedge \alpha$ with the evaluation map from $\mathfrak{g}E \otimes E$ to E. One then gets Formula (1) by a direct computation. □

Notice that, for G abelian, Formula (1) says that the curvature depends linearly on the connection. This is precisely what we have been using in our Yang-Mills presentation of electromagnetism in Lecture 1.

In local coordinates (x^i) of the base and for a trivialization (e_α) of the bundle, a connection ∇ can be expressed for a local section $s = s^\alpha e_\alpha$ of the bundle E as

$$\nabla s = \left(\frac{\partial s^\alpha}{\partial x^i} + \Gamma^\alpha_{i\,\beta} s^\beta \right) dx^i \otimes e_\alpha \ .$$

Notice, that a connection is naturally associated with any trivialization ϕ of the bundle. We denote it by ∇^ϕ. (Just consider a section as a function on the base with values in a fixed vector space by means of its components in the trivialization.) The local coefficients $(\Gamma^\alpha_{i\,\beta})$ of the connection ∇ are precisely the components of the difference of ∇ and the connection ∇^ϕ. By a direct calculation using Formula (1), and using the fact that the connection ∇^ϕ has vanishing curvature one gets the local expression of the curvature

$$R^\alpha_{ij\,\beta} = \left(\frac{\partial \Gamma^\alpha_{i\,\beta}}{\partial x^j} - \frac{\partial \Gamma^\alpha_{j\,\beta}}{\partial x^i} \right) + \Gamma^\alpha_{i\,\gamma} \Gamma^\gamma_{j\,\beta} - \Gamma^\alpha_{j\,\gamma} \Gamma^\gamma_{i\,\beta} \ .$$

(Here, we have assumed that the group G is linear so that we can write the Lie bracket as a commutator of linear maps.)

If we now give ourselves a Riemannian metric g on the base manifold M (hence, also a volume element v_g), we are in a position to define the Yang-Mills functional YM, as

$$YM(\nabla) = \frac{1}{2} \int_M |R^\nabla|^2 v_g$$

where $|\ |$ stands for a metric norm on $\Lambda^2 T^*M \otimes \mathcal{g} E$ that we take to be $g \otimes k$ for an adG-invariant metric k on $\mathcal{g}$, the opposite of the Killing form for example if G is semi-simple.

Notice, that in order that the integral makes sense, we have either to assume that the manifold M is compact or that the field R^∇ decreases sufficiently fast at infinity. The theory we will be developing always assumes that these assumptions are met, i.e., that the functional is finite.

The functional as we defined it is sometimes called the Euclidean Yang-Mills functional to emphasize that one works with a metric of elliptic signature. (This fact is connected to current methods used in quantification, an interesting point on which we have no time, nor the qualification to elaborate.)

The local expression of the integrand of YM in local coordinates (x^i) on the base, and bundle coordinates (E^α) is

$$- g^{ik} g^{jl} R_{ij}{}^\alpha{}_\beta R_{kl}{}^\beta{}_\alpha \sqrt{\det(g_{mn})} \; dx^1 \ldots dx^n .$$

(Here, (g^{ij}) denotes the inverse matrix of the local expression (g_{ij}) of the metric g, and we assumed G to be a classical linear group and used the trace expression of the opposite of the Killing form as fibre metric, omitting a normalization factor.)

If one replaces the metric g by a conformally related metric $\lambda^2 g$, then the integrand gets multiplied by λ^{n-4}. Therefore, it is precisely in dimension 4 that it is conformally invariant. Physicists like to refer to this phenomenon as a scaling invariance having in mind the transformations which in simple spaces such as $\mathbb{R}^4$ or S^4 preserve the standard conformal class. This property is crucial for many of the arguments that we present later (cf. Lecture III).

2. The first variation formula.

Recall that we are working on a fixed Riemannian manifold and considering only connections for which the Yang-Mills functional is finite. We now accomplish a step that is basic for any variational problem, establishing the first variation formula (cf. the Lecture by K. Grove for analogous formulae concerning the length and the energy of curves).

First variation formula. If $(\nabla^t)_{t \in I}$ is a variation of the connection ∇ so that $\frac{d \nabla^t}{dt}\big|_{t=0} = A$, then the first variation of the Yang-Mills functional reads

(2) $$\frac{d}{dt} YM(\nabla^t)\big|_{t=0} = \int_M \left((d^\nabla)^* R^\nabla, A \right) v_g ,$$

where $(d^\nabla)^*$ denotes the adjoint of d^∇.

Proof. It is quite straightforward on the basis of Formula (1).

Since we are computing the first derivative of a function we can replace the curve $(\nabla^t)_{t \in I}$ through ∇ in the affine space $\mathcal{CE}$ by the curve $\nabla + tA$ which is equivalent to it up to first order. The quadratic term in (1) drops being of second order.

$$\frac{d}{dt} R^{\nabla + tA} \Big|_{t=0} = d^\nabla A \quad.$$

Formula (2) then follows after one integrates by parts, therefore introducing the adjoint operator. □

The first variation formula says that, at a connection ∇, $(d^\nabla)^* R^\nabla$ is the L^2-gradient of the functional. It is the Yang-Mills current when one considers the case of a non empty space. Hence, the Euler-Lagrange equation for YM, called the system of Yang-Mills equations, reads

$$d^\nabla {}^* R^\nabla = 0 \quad.$$

Notice that these field equations have the peculiarity of involving the potential also in the operator which is applied to the field. This adds to the non-linearity of the system which is already present because of the field expression when the structure group is non abelian. These phenomena disappear when G is abelian, e.g., in the case of electromagnetism for which $G = U_1$, hence the linearity of Maxwell's equations as we have seen before.

Notice also, that the base metric enters the Yang-Mills system via the adjoint operation on d^∇.

A critical point of YM is called a Yang-Mills connection. (Notice here the difference in terminology between mathematicians and physicists, since for physicists a Yang-Mills potential, or a Yang-Mills field, refers only to the fact that one is working in a Yang-Mills setting.)

On a given G-bundle over a given Riemannian manifold (M,g) the main question is to describe all Yang-Mills connections.

In local coordinates, the system of Yang-Mills equations can be written as follows

$$0 = - g^{ij} \left\{ \frac{\partial^2 \Gamma_{k\ \beta}^{\alpha}}{\partial x^i \partial x^j} + \frac{\partial^2 \Gamma_{i\ \beta}^{\alpha}}{\partial x^j \partial x^k} \right\} -$$

$$- 2g^{ij} \left[\Gamma_i, \frac{\partial \Gamma_k}{\partial x^j} \right]_\beta^\alpha + g^{ij} \left[\Gamma_i, \frac{\partial \Gamma_j}{\partial x^k} \right]_\beta^\alpha - g^{ij} \left[\frac{\partial \Gamma_j}{\partial x^i}, \Gamma_k \right]_\beta^\alpha$$

$$- g^{ij} \left[\Gamma_i, [\Gamma_j, \Gamma_k] \right]_\beta^\alpha .$$

It is clearly a second order system of partial differential equations in the local expression of the connection. The second order terms are linear, the first order terms linear in the Γ's and their derivatives, the zeroth order terms being cubic in the coefficients.

Now, notice that for any connection ∇ on any bundle the differential Bianchi identity says that

$$d^\nabla R^\nabla = 0 .$$

(In fact, this directly follows from the definition of R^∇ that we gave. Indeed, R^∇ being a derivation, for a section s of E,
$(d^\nabla R^\nabla)(s) = d^\nabla(R^\nabla(s)) - R^\nabla(d^\nabla s) = d^\nabla \circ (d^\nabla \circ d^\nabla s) - (d^\nabla \circ d^\nabla) \circ d^\nabla s = 0$.)

Since the curvature field of a Yang-Mills connection, a so-called Yang-Mills field, is both closed and coclosed, Yang-Mills theory is often called a non-linear Hodge theory.

3. **The gauge group.**

We must introduce the invariance group of the functional which shapes up the theory. The gauge group GE is the infinite dimensional group of sections of the automorphism bundle GE, the gauge transformations. It acts naturally on CE in the following way. For a section s of E, an element γ of GE, and an element ∇ of CE, one sets

$$\nabla^\gamma s = \gamma(\nabla(\gamma^{-1}(s))) .$$

Then, one easily sees that $R^{\nabla^\gamma} = \gamma \circ R^\nabla \circ \gamma^{-1}$. Therefore, the action of G on E being assumed orthogonal,

$$YM(\nabla^\gamma) = YM(\nabla) \quad ,$$

i.e. the Yang-Mills functional is invariant under the gauge group. One can then consider YM as a function on the quotient space CE/GE, a space which has a richer topology than CE.

Notice here again some discrepancies of terminology between mathematicians and physicists. Sometimes in the physics literature the gauge group is called the local invariance group, and the structure group the gauge group.

From the global point of view, one can make the following remark. If ∇ is a Yang-Mills connection, so are the ∇^γ's for all gauge transformations γ. Therefore, all tangent vectors to CE obtained as $\frac{d}{dt}\nabla^{\gamma_t}|_{t=0}$ at a Yang-Mills connection ∇ for paths $t \mapsto \gamma_t$ in E form the infinite dimensional space $d^\nabla \Omega^0(M, gE)$ lying in the kernel of the linearized Yang-Mills equation. Hence, these equations cannot be elliptic. In fact, the tangent directions that we exhibited are the only characteristic directions of the principal symbol. In some sense, one can say that, on the space CE/GE of inequivalent potentials, the system of Yang-Mills equations is elliptic.

Of course, this difficulty is not present if one considers this system as an equation on the curvature. This is the point of view that we adopt in the next section.

If one is to deal with the system of Yang-Mills equations as a system having the connection as unknown, one has to break the invariance under the gauge group by working in an appropriately chosen gauge. To see this more specifically, consider how a connection $\nabla = \nabla^\phi + A$ changes under a gauge transformation τ. We get

$$\nabla = \nabla^\phi + A^\gamma = \nabla^\phi + \gamma^{-1}(d^{\nabla^\phi}\gamma) + \gamma^{-1} \circ A \circ \gamma \quad ,$$

in which A^γ is the local expression of the connection ∇^γ. In order to apply analytical techniques, one has to choose the gauge transformation properly in order to make the system elliptic. (See the lectures by A. Bahri for details on that.)

4. Yang-Mills equations as an elliptic system in the curvature.

This point of view works only to show that certain Yang-Mills fields do not exist. Thanks to it, one can show that on some specific base spaces, some Yang-Mills fields are isolated. The key technique is the use of a Bochner-Weitzenböck formula for the Yang-Mills field R^∇, which is a $\mathfrak{g}E$-valued harmonic 2-form as we mentioned in the previous section. Such a tool is quite classical in Riemannian geometry. The main lemma is the following identity relating the Hodge-de Rham Laplacian $d^\nabla(d^\nabla)^* + (d^\nabla)^* d^\nabla$ to the rough Laplacian $\nabla^*\nabla$.

Bochner-Weitzenböck formula. For any E-valued differential 2-form α, and tangent vectors X, and Y,

$$(3) \quad \left((d^\nabla(d^\nabla)^* + (d^\nabla)^* d^\nabla)\alpha\right)_{X,Y} = \nabla^*\nabla\alpha_{X,Y} + \alpha \circ c(R^g)_{X,Y} +$$

$$+ \sum_{i=1}^{n} \left\{ \left[R^\nabla_{e_i,X}, \alpha_{e_i,Y}\right] - \left[R^\nabla_{e_i,Y}, \alpha_{e_i,X}\right] \right\},$$

where $c(R^g) : \Lambda^2 T^*M \to \Lambda^2 T^*M$ is a linear map depending linearly on the Riemannian curvature of the metric g (which for the standard metric on the sphere S^n is just $2(n-2)\,\text{Id}$).

Details on the proof can be found in [B-L] page 200.

We conclude this lecture by a gap theorem which can be refined in dimensions 3 and 4. We reserve the discussion of this point to Lecture III.

Gap Theorem (cf. [B-L]). Any Yang-Mills field R^∇ on the standard sphere S^n with $n \geq 5$ which satisfies the pointwise estimate

$$2|R^\nabla|^2 \leq \binom{n}{2}$$

vanishes identically.

Proof. We apply the Bochner-Weitzenböck Formula (3) to the Yang-Mills field R^∇. The left-hand side is identically zero since the 2-form R^∇ is harmonic. By integrating it against itself over the sphere, we get

$$\int_M |\nabla R^\nabla|^2 v_g = \int_M \sum_{i,j,k,=1}^n \left(\left[R^\nabla_{e_i,e_j}, R^\nabla_{e_j,e_k} \right], R^\nabla_{e_k,e_i} \right) v_g ,$$

which gives a contradiction if one proves that the right hand side is non positive. This is indeed the case as the next lemma shows. (The proof, a purely algebraic fact of independent interest, can be found in [B-L] page 206.)

Lemma. If $2|R^\nabla|^2 \leq \binom{n}{2}$ and $n \geq 5$, then

$$\sum_{i,j,k=1}^n \left(\left[R^\nabla_{e_i,e_j}, R^\nabla_{e_j,e_k} \right], R^\nabla_{e_k,e_i} \right) < 2(n-2) |R^\nabla|^2 .$$

REFERENCE

[B-L] J.P. BOURGUIGNON, H.B. LAWSON Jr., Stability and isolation phenomena for Yang-Mills fields, Commun. Math. Phys. 79 (1981), 189-230.

Lecture III. SPECIAL FEATURES OF 4-DIMENSIONAL YANG MILLS THEORY.

When the dimension of the base space M is 4, the typical case of interest to physicists, special features appear. One of them is in disguise the fact that the special orthogonal group SO_4 is not simple, a well known fact. Another one is conformal invariance of the Yang-Mills functional.

(An excellent general reference for the material of this Lecture, and other things, is [A-H-S].)

1. Self-duality.

To begin with, we introduce the Hodge map $*$ which, on an oriented n-dimensional Euclidean vector space (whose metric is denoted by g), attaches to a k-form α an (n-k)-form $*\alpha$ in the following way. For any k-form β , one sets

$$*\alpha \wedge \beta = (\alpha,\beta) \cdot v_g .$$

(Here, we have used the orientation to view the volume element v_g as an n-form.) One can also say that, if (e_i) is a positive orthonormal basis and I a multiindex, $*(e_I) = e_J$ where the (n-k)-multi-index J is such that $\{I,J\}$ is an even permutation of $\{1,\ldots,n\}$. It is then clear that on k-forms $* \circ * = (-1)^{k(n-k)}$ and that, on a 2k-dimensional vector space, $*$ as a map on k-forms depends only on the conformal class of the metric. (Indeed, for the metric $\lambda^{-2} g$, the basis $(\lambda^{-1} e_i)$ is orthonormal, hence $*(\lambda^{-k} e_I) = \lambda^{-k} e_J$ if $|I| = |J| = k$.)

On a 4-dimensional vector space $*$ is an involution of the 6-dimensional space $\wedge^2 V^*$. We decompose the space $\wedge^2 V^*$ into the (+1)-eigenspace $\wedge^+ V^*$ and the (-1)-eigenspace $\wedge^- V^*$ of $*$. Elements of $\wedge^+ V^*$ and $\wedge^- V^*$ are respectively called self-dual and antiself-dual 2-forms.

If (e_i) is a positive orthonormal basis of V with dual basis (ε^i) , then $(\varepsilon^1 \wedge \varepsilon^2 + \varepsilon^3 \wedge \varepsilon^4, \varepsilon^1 \wedge \varepsilon^3 + \varepsilon^4 \wedge \varepsilon^2, \varepsilon^1 \wedge \varepsilon^4 + \varepsilon^2 \wedge \varepsilon^3)$ and respectively $(\varepsilon^1 \wedge \varepsilon^2 - \varepsilon^3 \wedge \varepsilon^4, \varepsilon^1 \wedge \varepsilon^3 - \varepsilon^4 \wedge \varepsilon^2, \varepsilon^1 \wedge \varepsilon^4 - \varepsilon^2 \wedge \varepsilon^3)$ are bases of $\wedge^+ V^*$ and $\wedge^- V^*$.

Notice that, if the metric g is Lorentzian, $*$ becomes a complex structure on $\wedge^2 V^*$. In that case, the space remains irreducible over the real numbers.

These constructions can of course be performed pointwise on a 4-dimensional oriented Riemannian manifold M. Since the curvature is a 2-form, we shall consider its decomposition

$$R^\nabla = R^\nabla_+ + R^\nabla_-$$

into its self-dual and antiself-dual parts.

Another useful point of view is to consider the Lie algebra of $SO(V,g)$. It can be identified with the space of 2-forms on V. It is easy to see that the elements in $\wedge^+ V$ (resp. in $\wedge^- V$) correspond to linear maps which are multiple of complex structures which induce (resp. reverse) the fixed orientation. It is then an easy check to verify that these linear maps commute with one another. In this way we described the two ideals of the Lie algebra of $SO(V,g)$, hence showing that it can be decomposed as the direct sum of two copies of the Lie algebra of SU_2. The group SO_4 itself does not decompose, although dimension 4 is the only dimension in which the orthogonal group is not simple. We come back to this point in Section 2 using quaternions while introducing spinors.

Some other algebraic facts important for Yang-Mills theory follow from these considerations. We mention here one which has important applications while considering stable Yang-Mills fields (cf. Lecture IV).

Lemma. The space of traceless symmetric 2-tensors of an oriented 4-dimensional Euclidean vector space is isomorphic as an SO_4-module with the tensor product of self-dual 2-forms by antiself-dual 2-forms.

Proof. Notice first that the space $S^2_o V$ of traceless symmetric tensors is 9-dimensional as is the tensor product $\wedge^+ V \otimes \wedge^- V$.

The description of self-dual (resp. antiself-dual) 2-forms as linear maps suggests that with the tensor product of a self-dual 2-form α and of an antiself-dual 2-form β we associate their composition $\alpha \circ \beta$ as linear maps. We already noticed that they commute, hence $\alpha \circ \beta$ is symmetric. Its trace, the inner product of α and β, vanishes since
$(\alpha,\beta) = (*\alpha,\beta) = (\alpha,*\beta) = -(\alpha,\beta)$.

On the other hand, this map is clearly equivariant under the action of $SO(V,g)$. Since it is not the zero map, and since the space $S^2_o V$ is an irreducible representation space of $SO(V,g)$, it must be an isomorphism. (If it were not, its image would be an invariant subspace, a contradiction.)

□

In Lecture IV, we shall apply this lemma to the traceless symmetric 2-tensor (with values in gE) $\sum_{i=1}^{4} [R^{\nabla}_{+e_i,\cdot}, R^{\nabla}_{-e_i,\cdot}]$ (here, (e_i) is a positive orthonormal basis) which is another expression of the composition of a self-dual and an antiself-dual 2-form.

To come back to the system of Yang-Mills equations, we first notice that the adjoint d^* of the differential d on (ordinary or vector-valued) 2-forms has the following expression in terms of the Hodge map

$$d^* = - * \circ d \circ * \ .$$

The system of Yang-Mills equations can therefore be rewritten

(4) $$d^{\nabla}(*R^{\nabla}) = 0 \ .$$

In dimension 4, this exhibits its only dependence on the conformal class of the base metric, via the Hodge map $*$ on 2-forms. It therefore follows from the differential Bianchi identity that a self-dual or an antiself-dual curvature is automatically a Yang-Mills field, a crucial property that we are going to discuss again in Sections 3 and 4. Notice that the self-duality equations are algebraic in the curvature, hence first order in the connection.

Another consequence of the formulation (4) of the Yang-Mills equations and of the Bianchi identity is the following Proposition whose proof is straightforward and left to the reader.

<u>Proposition.</u> The following statements are equivalent:

i) the field R^{∇} satisfies Yang-Mills equations;

ii) the self-dual part R^{∇}_+ of the field is closed;

iii) the antiself-dual part R^{∇}_- of the field is closed.

2. An algebraic introduction to spinors in dimension 4.

It is now clear to geometers that the notion of spinors is going to play an important role in the study of geometric objects. Of course, this notion revealed itself crucial for physics for quite a while. It is not our purpose here to define spinors in full generality, but rather to see how they help understanding some of the special properties of dimension four. (A possible reference on the algebraic side of the theory of spinors is

[A-B-S], one more adapted to our purposes is the appendix of [Be].)

The notion of spinors is meaningful on any Euclidean vector space (V,g). (In other words, it depends on the choice of a metric.) The subtle point about spinors is that it is not easy to define them directly. One catches them through their algebra of automorphisms, the Clifford algebra $Cl(M,g)$ attached to the metric g. It is defined as the quotient of the tensor algebra by the ideal generated by elements of the form $X \otimes X + g(X,X) 1$ for vectors x in V. It is universal for morphisms Φ of algebras with units which satisfy $(\Phi(X))^2 = - g(X,X) 1$. Hence, for vectors X and Y of V, the following relation holds

$$X.Y + Y.X = -2 g(X,Y) 1 .$$

(Here, we used . to denote the product in the Clifford algebra, the so-called Clifford multiplication.)

It is therefore easy to see that $Cl(M,g)$ is 2^p-dimensional. (For an orthonormal basis (e_i) of V, a basis of it is given by elements (e_I) where I is a multi-index. Indeed, if follows from the commutation relations that any decomposable element in the basis elements can be put in this form.)

The structure of this algebra depends on the dimension, exhibiting an 8-fold periodicity over the real numbers, and a 2-fold periodicity over the complex numbers. In even dimensions, the complexified Clifford algebra $\mathbb{C}Cl(V,g)$ is a simple algebra over $\mathbb{C}$, i.e., is isomorphic to the full linear algebra of a complex vector space, the space $\Sigma(V,g)$ of spinors.

The Clifford algebra retains a $\mathbb{Z}_2$-grading from the tensor algebra (because the ideal by which one divides has only elements of even orders). One has $Cl(V,g) = Cl^+(V,g) \oplus Cl^-(V,g)$. In even dimensions, it is interesting to consider a further decomposition of the space of spinors into their positive and negative parts,

$$\Sigma(V,g) = \Sigma^+(V,g) \oplus \Sigma^-(V,g) ,$$

in such a way that

$$Cl^+(V,g) = \Sigma^{+*} \otimes \Sigma^+ \oplus \Sigma^{-*} \otimes \Sigma^-$$

and

$$Cl^-(V,g) = \Sigma^{+*} \otimes \Sigma^- \oplus \Sigma^{-*} \otimes \Sigma^+ .$$

We now detail the situation in dimension 4. The Clifford algebra $Cl(\mathbb{R}^4,e)$ (where e is the standard Euclidean metric on $\mathbb{R}^4$) that we denote by Cl_4

can be identified with HGl_2. The structure of $\mathbb{C}Cl_4$ is isomorphic to $\mathbb{C}Gl_4$. It is usual to speak mainly of the complexified spinors which in this case form a 4-dimensional complex vector space, that we denote by Σ_4, the only irreducible $\mathbb{C}Cl_4$-module in fact. This space Σ_4 is also the only irreducible representation space of the group $Spin_4$, the universal cover of SO_4. In general dimensions, it is possible to give a description of the group $Spin_n$ as a subgroup of the multiplicative group of the Clifford algebra. For our case, we shall use the quaternionic formalism to get hold of it. If we identify $\mathbb{R}^4$ with $\mathbb{H}$, for unit quaternions q, and q', we get an orthogonal map $x \longrightarrow qx q'^{-1}$. Clearly, multiplying at the same time q, and q' by -1 does not change the transformation of $\mathbb{R}^4$ that one gets. Since it is well known that the group of unit quaternions can be identified with the group SU_2 (just think of $\mathbb{H}$ as $\mathbb{C} \oplus \mathbb{C}$), in this way, one defines a 2-to-1 map from $SU_2 \times SU_2$ to SO_4, hence the possibility of identifying $Spin_4$ with $SU_2 \times SU_2$. In this description, both the spaces of positive and negative spinors can be identified with $\mathbb{C}^2$ for the standard representation of SU_2, and we also obtained a description of $\mathbb{R}^4 \otimes \mathbb{C}$ as the tensor product $\Sigma^+ \otimes \Sigma^-$, a spinor description of vectors! Along the same lines, one can describe the space $\wedge^{\pm} \mathbb{R}^4 \otimes \mathbb{C}$ as $S^2 \Sigma^{\pm}$. These identifications are very useful while dealing with 4-dimensional objects.

3. A glance at characteristic classes.

The theory as we developed it so far is local, i.e., all the quantities that we used are defined in local terms. This is true also of the first variation formula by using local variations with compact support. In this section, we shall encounter the global notion of characteristic classes of bundles, which can be introduced in many different more or less equivalent ways. We are going to approach them while sticking as much as possible to the notions we have introduced so far at the expense of missing some of their very important properties. (We refer the reader to [M-S] for a systematic exposition.)

We start by the following proposition.

Proposition.

On any compact manifold M, the expression

$$8\pi^2 k = \int_M \{|R_+^\nabla|^2 - |R_-^\nabla|^2\} v_g$$

depends neither on the connection ∇ on the bundle E, nor on the Riemannian metric on the base manifold M. It is a topological invariant of the bundle $\pi : E \longrightarrow M$ which is quantized, i.e., equal to $4\pi^2 k(E)$ where the relative integer $k(E)$ is half the Pontryagin index of the bundle.

Proof. We first prove the independence of the integral expression from the metric. (This part is elementary algebra.) One simply notices that the expression $\{|R_+^\nabla|^2 - |R_-^\nabla|^2\} v_g$ can be rewritten as $\{(R_+^\nabla, *R_+^\nabla) + (R_-^\nabla, *R_-^\nabla)\} v_g$, or else as $(R_+^\nabla + R_-^\nabla, *(R_+^\nabla + R_-^\nabla)) v_g$, and by the definition of the map $*$, $(R^\nabla \wedge R^\nabla)$ where the symbol ( ) now only represents the inner product on the Lie algebra $\mathfrak{g}$.

The integrand has been made free of any reference to the base metric.

The independence of k from the connection is usually referred to as Chern-Weil theory. We start from the expression $(R^\nabla \wedge R^\nabla)$, and we study how it varies when the connection varies. It is convenient to evaluate the first variation of this expression for an infinitesimal change A of the connection. From Formula (1) follows that the first variation of k is given by

$$\int_M (R^\nabla \wedge d^\nabla A) .$$

Now, notice that the following formula holds

$$d(R^\nabla \wedge A) = (d^\nabla R^\nabla \wedge A) + (R^\nabla \wedge d^\nabla A) .$$

(It is a slight extension to vector-valued forms of the usual derivation property of the exterior differential using the fact that the fiber inner product denoted by ( ) is parallel with respect to ∇.) The result is now clear since $d^\nabla R^\nabla = 0$ by the differential Bianchi identity.

At this point, we proved that the number k depends only on the bundle $\pi : E \longrightarrow M$. To prove that it is indeed an integer requires more work, in particular that one connects it with the integral cohomology of some space. We refer to [M-S] for that.

The number $k(E)$ is often called the instanton number or the topological quantum number by physicists. When $M = S^4$ and the group G simple, this number k is known to classify the bundle. (This follows from the possibility of covering the sphere by two charts over which the bundle must be trivial, reducing the problem of its classification to the description of the gluing map into the structure group B along the equatorial 3-sphere, hence to homotopy classes of maps from S^3 into G, cf. [Sd] for details). This terminology is suggested by the fact that certain solutions to the Yang-Mills equations can be interpreted as pseudoparticles, and that k counts them.

A basic consequence of this fact is to provide us with a topological lower bound for the Yang-Mills functional. Indeed, by noticing that
$|R^\nabla|^2 = |R^\nabla_+|^2 + |R^\nabla_-|^2$, one obtains

$$4\pi^2 \, |k(E)| \leq YM(\nabla)$$

with equality if and only if the connection ∇ is self-dual or antiself-dual. Hence, the self-duality equations give automatically absolute minima of the functional, a phenomenon which should be put into correspondence with the fact that holomorphic maps between Kähler manifolds minimize the energy in their homotopy classes (cf. F. Burstall's lectures).

4. Comments on self-dual connections over S^4.

That these equations can actually be satisfied on any bundle with its standard conformal structure was proved for the first time by 't Hooft. A complete description of all these solutions was obtained by M.F. Atiyah, V.G. Drinfeld, N. Hitchin, and Y.I. Manin using the Penrose twistor construction (cf. J. Rawnsley's lectures).

A differential geometric picture of the self-dual connections on the simplest non-trivial SU_2-bundle over (S^4,c) can be given as follows. The SU_2-bundles with $k = 1$ and $k = -1$ on S^4 are respectively the positive and negative spinor bundles. (Indeed, $Spin_4 = SU_2 \times SU_2$). The total spaces of the principal spinor bundles can be identified with the sphere S^7 fibered over the quaternionic projective line $\mathbb{H}P^1 = S^4$. The connections induced on these bundles by the Levi-Civita connections of the family parametrized by the ball B^5 of conformally equivalent constant curvature metrices on S^4 are all self-dual (cf. [A-H-S]). Using this description via conformal transformations, one can attach a center and a scale to these solutions, making them look like pseudoparticles, hence the name instantons given to them. (Explicit calculations concerning this construction can be found in [F-U] page 99-105, or in [A] Chapter II.)

Thanks to the decomposition of the curvature into its self-dual and antiself-dual parts, and to the existence of the topological conserved quantity k(E) , up to a constant, one can view the Yang-Mills functional as

$$\frac{1}{2} \int_M |R^\nabla_+|^2 \, v_g \quad \text{or} \quad \frac{1}{2} \int_M |R^\nabla_-|^2 \, v_g \ .$$

This together with the Proposition given in Section 1 about the harmonicity of the positive and negative parts of a Yang-Mills field allows to obtain refinements of the gap theorem given at the end of Lecture II.

Theorem ([D-M]). Any Yang-Mills connection ∇ on the sphere S^4 which satisfies $4\pi^2|k| \leq YM(\nabla) \leq 4\pi^2(|k|+1)$ is self-dual if $k \geq 0$ and antiself-dual if $k \leq 0$.

The proof is obtained by applying the Bochner-Weitzenböck formula to the harmonic 2-form R^∇_+ (or R^∇_- depending on the sign of k), and instead of getting estimates using the uniform norm one applies Hölder's inequalities and then Sobolev inequalities with the best constants. The advantage of such a theorem over the one given earlier is that it gives a neighbourhood of self-dual fields in the space of connections containing no other critical point in a weaker topology, in particular in one which is directly connected with the variational problem.

REFERENCES

[A]   M.F. ATIYAH, Lectures on Yang-Mills fields, Lezioni Fermi Ac. Naz. dei Lincei Scuola Norm. Sup. Pisa (1979).

[A-B-S] M.F. ATIYAH, R. BOTT, A. SHAPIRO, Clifford modules, Topology 3 (1964), 3-38.

[A-H-S] M.F. ATIYAH, N. HITCHIN, I.M. SINGER, Self-duality in 4-dimensional Riemannian geometry, Proc. Roy. Soc. London A, 362 (1978), 425-461.

[B-L]   J.P. BOURGUIGNON, H.B. LAWSON Jr., Stability and isolation phenomena for Yang-Mills fields, Commun. Math. Phys. 79 (1981), 189-230.

[Be]   A.L. BESSE, Géométrie riemannienne de dimension 4, Séminaire, CEDIC-Nathan, Paris (1981).

[D-H]   J. DODZIUK, M. MIN'OO, An L^2-isolation theorem for Yang-Mills fields over complete manifolds, Compositio Math. 47 (1982), 165-169.

[F-U]   D. FREED, K.K. UHLENBECK, Instantons and four-manifolds, Math. Sci. Res. Inst. Publ. no.1, Springer, (1984).

[M-S]   J.W. MILNOR, J. STASHEFF, Characteristic classes, Ann. Math. Studies no. 76, Princeton University Press, Princeton (1974).

[Sd]   N. STEENROD, The topology of fibre bundles, Princeton University Press, Princeton (1951).

Lecture IV. THE SECOND VARIATION AND STABLE YANG-MILLS FIELDS

For any geometric variational problem, to know the second variation formula at a critical point is always a very important piece of information. Yang-Mills theory is no exception. Critical points for which the second variation is non-negative, like local minima, are expected to have some stability properties, hence play a key role for obvious physical reasons. On spaces which present enough symmetries, it is possible to prove that stable Yang-Mills fields (stability is a priori a global differential constraint on them) verify some pointwise algebraic conditions. This is the content of the stability theorems that we present in this lecture.

1. The second variation formula.

We establish the second variation formula at a Yang-Mills connection ∇ , since at a non-critical point the second derivative of the functional does not make sense unless we choose a connection on the space $\mathcal{C}_E$.

<u>Second variation formula.</u> Let ∇ be a Yang-Mills connection over a bundle $\pi : E \longrightarrow M$. For any variation $(\nabla^t)_{t \in I}$ of ∇ for which $\frac{d}{dt}\nabla^t\big|_{t=0} = A$, we have

(5) $$\frac{d^2}{dt^2} YM(\nabla^t)\big|_{t=0} = \int_M \left\{ |d^\nabla A|^2 + \left(\sum_{i=1}^n \left[R^\nabla_{e_i}, \, . \, , A \right], A_{e_i} \right) \right\} v_g .$$

<u>Proof.</u> It is quite straightforward. For simplicity, we replace the variation $(\nabla^t)_{t \in I}$ by its second jet $\nabla + tA + \frac{t^2}{2} B$, and as usual when one evaluates a second derivative at a critical point, we expect B to disappear in the final formula.

The right hand side of Formula (5) is obtained directly form Formula (1) by retaining the terms of degree 2 in t after one transforms $([A \wedge A], R^\nabla)$ into the term given there because the inner product on the Lie algebra has been chosen biinvariant. The term containing $d^\nabla B$ indeed disappears since it is integrated against the Yang-Mills field R^∇ .

It is interesting to compare the second variation formula that we just described with the index form connected with the second variation of arc length in K. Grove's lectures. They have a similar structure. Both of them contain one term involving some part of L^2-norm of the derivative of the variation field, and one term involving linearly the curvature of the critical object. There is though one crucial difference. In the Yang-Mills setting, the term $d^\nabla A$ containing the derivative of the variation field is not elliptic. This reflects the presence of the infinite dimensional invariance group, the gauge group GE. To state this more precisely, it is convenient to introduce the Jacobi operator

$$J^\nabla(A) = (d^\nabla)^* d^\nabla A + \sum_{i=1}^{n}\left[R^\nabla_{e_i}, \cdot\,, A_{e_i}\right]$$

defined in such a way that

$$\int_M (J^\nabla(A), A)\, v_g = \frac{d^2}{dt^2}\, YM(\nabla + tA)\Big|_{t=0} \,.$$

Because of the invariance of the functional YM under GE, it is indeed true that the Jacobi operator annihilates any tangent vector to the orbit of the connection ∇ under the gauge group. It is an easy exercice to verify that any such vector can be written $d^\nabla \Gamma$ for Γ in $\Omega^0(M, gE)$. (For that, one only has to compute the derivative at the identity of the formula giving the modification of a connection after a gauge transformation, formula which is given in Lecture II). So, we have the tautological identity $J^\nabla(d^\nabla \Gamma) = 0$, which can be used for example in order to check the second variation formula.

At this point, it may be interesting to make a comment on the geometric structure of the space of connections. Using the fact that the operator d^∇ acting on $\Omega^0(M, gE)$ is elliptic in a generalized sense (its symbol is injective), one gets a direct sum decomposition

$$T_\nabla CE = \Omega^1(M, gE) = \operatorname{Im} d^\nabla \oplus \operatorname{Ker}(d^\nabla)^*$$

where $\operatorname{Ker} d^{\nabla *}$ in this generalized situation is infinite dimensional. This suggests that to work transversally to the action by the gauge group (what one should do when one is really interested in geometric properties), one should consider variations whose first jets lie in the space $\operatorname{Ker} d^{\nabla *}$.

While doing that to the Jacobi operator J^∇ one can substitute a modified Jacobi operator $\tilde{J}^\nabla = J^\nabla + d^\nabla \circ (d^\nabla)*$, whose leading term is now obviously elliptic.

What we just described is one step in the direction of defining the topology of the space of orbits. One can in fact prove a slice theorem saying that appropriate neighbourhoods in the transversal space $\text{Ker } d^\nabla *$ cut all orbits near that of ∇. (A precise statement can be found in [F-U] page 57.)

By definition, the second variation formula detects the behaviour of the functional up to second order at a critical point. A Yang-Mills connection ∇ is said to be stable if the second variation at ∇ is non-negative, in other words if the Jacobi operator J^∇ is a non-negative second order differential operator. Typical examples of stable Yang-Mills connections are given by local minima of the Yang-Mills functional.

Since the modified Jacobi operator (which carries all necessary geometric information) is elliptic with principal part equal to the Hodge-de Rham Laplacian, there is at most finitely many negative eigenvalues, i.e., any subspace on which J^∇ restricts to a negative operator is finite dimensional. Using an analogy with finite dimensional Morse theory (cf [M]), one can say that any critical point of the Yang-Mills functional has finite index. We already pointed out that the kernel of the Jacobi operator J^∇ is necessarily infinite dimensional.

2. Special variations : the enlarged gauge group.

The proper use of the second variation formula at a critical point of a geometric functional is to evaluate it on some properly chosen tangent vectors in order to get more refined information on the critical point by taking advantage of its geometric (or algebraic) structure.

Invariance properties of the functional usually play a very important role at this stage. We give an example of this in the theory of geodesics (cf K. Grove's lectures). It is clear that isometries preserve the length functional defined on curves. Therefore, the image of a geodesic under an isometry remains a geodesic. Then, if X is an infinitesemal isometry (one usually says a Killing vector field), since the image of a geodesic γ under the flow of X is a variation of γ by geodesics, X is a Jacobi field along γ, i.e., lies in the kernel of the Jacobi operator for the length functional.

We shall use a similar idea in Yang-Mills theory in order to find interesting variation vectors at a Yang-Mills connection ∇. Recall that the gauge group GE is the group of transformations of the bundle which induce automorphisms in each fibre, and therefore projects onto the base manifold M as the identity map of M. Since the Yang-Mills set-up depends only on the Riemannian metric on M, it is obvious that if we now consider the group of automorphisms of E which project onto M as isometries, we again obtain an invariance group for the Yang-Mills functional, the enlarged gauge group $\widetilde{G}E$. When the dimension of M is 4, since the Yang-Mills functional is known to depend only on the conformal class of the base metric, we can even enlarge further the gauge group by considering automorphisms of E which project onto M as conformal transformations.

If $t \longmapsto \gamma_t$ is a curve through the origin in $\widetilde{G}E$, we are interested in knowing how it acts infinitesimally at a connection ∇. The variations we shall be interested in are those projecting onto non trivial vector fields on the base manifold. One can check that the infinitesimal action on the connection ∇ of the element of the infinitesimal gauge group determined by the ∇-horizontal lift in the bundle $GE \longrightarrow M$ of a vector field X on M is given by the E-valued differential 1-form $i_X R^\nabla$. (We leave this as an interesting exercise to the reader.)

Therefore, when X is a Killing field, or a conformal vector field when the dimension of M is 4, it is a tautological identity that

$$J^\nabla(i_X^\nabla R) = 0 .$$

To get interesting identities, one should consider geometric modifications of the variations we just considered. This is indeed what J. Simons did when he took infinitesimal variations of the connection of the form $i_X^\nabla R$ for X a conformal vector field in dimension $n \geq 5$ and found that

(5) $$J^\nabla(i_X^\nabla R) = (n-4) i_X^\nabla R .$$

(As we announced, the formula exhibits the fact that in dimension 4 those variations have to annihilate tautologically the Jacobi operator.)

When M is 4-dimensional, one has to introduce other types of variations. Since the self-dual and antiself-dual parts of a Yang-Mills field are harmonic 2-forms, it is natural to consider variations of the type $i_X R_+^\nabla$ or $i_X R_-^\nabla$ where X is a conformal vector field or even a Killing field. For those, the following identity holds

(6) $$J^\nabla(i_X R_+^\nabla) = \sum_{i=1}^{4} [R_{+_{e_i, X}}, R_{-_{e_i}}, .]$$

(where (e_i) is an orthonormal basis of the tangent space). Notice that in the right hand side appears precisely the algebraic quantity we considered in Lecture III which was unexpectedly symmetric.

These formulas follow from Weitzenböck formulas taking into account the pointwise identities satisfied by the special vector fields that we considered.

3. The basic stability theorem.

In this section, we consider stable Yang-Mills fields, and show that over some special manifolds they have very particular properties. Theorems of this type are quite classical in global differential geometry (cf the use of minimizing geodesics in the study of the length functional, e.g., in the proof of Myers' theorem in K. Grove's lecture notes).

We first mention J. Simons' theorem because it played an important role in the birth of the other stability theorems. He proved that there does not exist any stable Yang-Mills field over the standard sphere S^n ($n \geq 5$). The proof is based on Formula (5), and a method that we outline after having stated the basic stability theorem.

For that purpose, we need the notion of the self-dual (resp. antiself-dual) holonomy algebra a_+^∇. At any point m on M, we define a_+^∇ to be the algebra generated in gE_m by the elements $R_{+_{U,V}}^\nabla$ where U, and V are arbitrary vectors in $T_m M$. (Of course, R_-^∇ replaces R_+^∇ for the antiself-dual holonomy algebra.)

Basic Stability Theorem. If ∇ is a stable Yang-Mills connection over a compact orientable homogeneous Riemannian manifold of dimension 4, then

$$\left[a_+^\nabla, a_-^\nabla \right] = 0 .$$

Proof. We only give an outline of it. Details can be found in [B-L1] and [B-L2].

Since ∇ is a stable Yang-Mills connection, the Jacobi operator is non-negative. One applies it to the special variations $i_X R^\nabla_+$ (where X is a Killing vector field) mentioned in the previous section. The Second Variation Formula for those vector fields can be expressed in purely algebraic terms because of Formula (6). If one averages this variation on the unit sphere in the space of Killing fields, one still gets a non-negative quantity. If one interchanges integration over M and over this sphere, it becomes possible to evaluate this integral by choosing properly at each point of M an orthonormal basis of the space of Killing fields. (Notice that we use here the fact that in a Euclidean vector space averaging a quadratic functional over the sphere is the same as taking its trace in an orthonormal basis up to a universal factor.) Homogeneity of the manifold is used in order to find a basis of the tangent space at the point consisting of Killing fields at that point.

This integral of second variations turns out to be zero for algebraic reasons of the nature we mentioned in Lecture III. Therefore, this ensures us that for any Killing vector field X, $i_X R^\nabla_+$ lies in the kernel of the Jacobi operator. Hence, the right hand side of Formula (6) is identically zero over M. It takes again an application of the algebraic lemma given in Lecture III to finish the proof of the theorem.

$\square$

4. The self-duality theorems.

In this section, we try to derive more precise algebraic statements from the Basic Stability Theorem. This will lead us to make two important points. First of all, if the group is small enough, the commutation relation between the self-dual and the antiself-dual holonomy algebras are going to put strong constraints on these algebras, hence lead to the vanishing of one of them. Secondly, we shall meet the important phenomenon of reduction of connections.

What matters in the size of the group G is how big the centralizer of a non-trivial element can be. As we mentioned earlier, the group SO_4 is not simple, hence the possibility of a non trivial centralizer for non trivial elements. Therefore, we say that a structure group is small if it does not contain SO_4 as a proper subgroup. This is indeed the case for SU_2, SO_3, U_2, and SU_3. The case of SO_4 is simple enough to be dealt with (cf. [B-L1] for the notion of two-fold self-duality which is adapted to it). Notice that in the isolation theorems the nature of the group did not matter.

The second phenomenon is interesting since in the case of the group SU_2 for example it is tied to the topology of the base manifold M. Indeed, if an SU_2-connection ∇ reduces, i.e., if it takes its values in a lower dimensional Lie algebra bundle than gE, it means it is a U_1-connection. In this case, it is possible to understand the curvature as an ordinary exterior differential form. If ∇ is a Yang-Mills connection, this means that R^∇ is an ordinary harmonic 2-form. By the Hodge-de Rham theorem, this can only be the case if the second cohomology group of M does not vanish, hence the link to the topology. This fact will also be very important in the discussion of Donaldson's theorem (cf. A. Bahri's and C. Taubes' lectures).

We now state the main stability theorems for the group SU_2 before we give a brief outline of the proofs refering again to [B-L1] and [B-L2] for details, and more general statements.

Theorem. Any stable SU_2-Yang-Mills field over a compact orientable homogeneous Riemannian manifold of dimension 4 is either self-dual, or antiself-dual, or reduces to a U_1-field. In all cases, they are absolute minima of the Yang-Mills functional.

Outline of proof. Since $C = SU_2$, the centralizer of every non-trivial element is reduced to the line generated by this element. Then, at each point, either a_+^∇ or a_-^∇ is reduced to 0, or they are equal and 1-dimensional. This possibility is ruled out by the Bochner-Weitzenböck formula applied to R^∇ unless it reduces to a U_1-field. In the first situation, on an open set one of the two algebras a_+^∇ and a_-^∇ is reduced to zero, hence on all of M since the curvature R^∇ is a harmonic field, and behaves like an analytic function. □

Corollary. Any stable SU_2-Yang-Mills field over the standard sphere S^4 is either self-dual or antiself-dual.

REFERENCES

[B-L1] J.P. BOURGUIGNON, H.B. LAWSON Jr., Stability and isolation phenomena for Yang-Mills fields, Commun. Math. Phys. 79 (1981), 189-230.

[B-L2] J.P. BOURGUIGNON, H.B. LAWSON Jr., Yang-Mills theory : its physical origin and differential geometric aspects, in Seminar on Differential Geometry ed. by S.T. Yau, Ann. Math. Studies n° 102, Princeton University Press., Princeton (1982).

[F-U]   D. FREED, K.K. UHLENBECK, Instantons and four-manifolds, Math. Sci. Res. Inst. Publications 1 (1984), Springer.

[M]   J. MILNOR, Morse theory, Ann. Math. Studies n° 51, Princeton University Press., Princeton (1963).

TWISTOR METHODS FOR HARMONIC MAPS

by

Francis E. Burstall

Contents
- I   Introduction and Overview
- II Construction of 2-dim σ-models
- III Twistor spaces
- IV Twistor lifts
- V   Applications

I. Introduction and Overview

A. Harmonic maps are solutions of a natural variational problem in Differential Geometry. Let us begin by recalling the relevant definitions.

So let $\phi : (M,g) \to (N,h)$ be a C^∞ map of C^∞ Riemannian manifolds (henceforth all ingredients will be assumed C^∞ without comment). The *energy*, $E(\phi)$ of ϕ is given by

$$E(\phi) = \tfrac{1}{2}\int_M |d\phi|^2 * 1_M$$

where the norm in the integrand is the Hilbert-Schmidt (tensor product) norm on $T^*M \otimes \phi^{-1}TN$.

A map ϕ is said to be *harmonic* if it extremises the energy on all compact sub-domains of M. Thus ϕ is harmonic if and only if it is a solution of the associated Euler-Lagrange equations:

$$\tau_\phi = \text{Trace}\nabla d\phi = 0$$

where ∇ is the connection on $T^*M \otimes \phi^{-1}TN$ induced by the Levi-Civita connections on M and N. The Euler-Lagrange operator τ is called the *tension field*.

In local co-ordinates, $(x_1,\ldots,x_m)$ on M, $(y_1,\ldots,y_n)$ on N

$$\tau_\phi^\alpha = g^{ij}\left(\frac{\partial^2\phi^\alpha}{\partial x_i \partial x_j} - \frac{\partial\phi^\alpha}{\partial x_k}{}^M\Gamma_{ij}^k + {}^N\Gamma_{\beta\gamma}^\alpha(\phi)\frac{\partial\phi^\beta}{\partial x_i}\frac{\partial\phi^\gamma}{\partial x_j}\right)$$

where ${}^M\Gamma_{ij}^k$, ${}^N\Gamma_{\alpha\beta}^\gamma$ are the Christoffel symbols on M and N respectively. Thus harmonic maps are locally solutions of a system of semi-linear elliptic partial differential equations.

Harmonic maps occur in many different situations in Geometry and Physics, for example:

i. If $N = \mathbb{R}$, τ is just the Laplace-Baltrami operator on (M,g) and harmonic maps are just harmonic functions.

ii. If dim $M = 1$ harmonic maps are precisely geodesics parameterised by (an affine transformation of) arc length.

iii. If dim $M = 2$ and ϕ is weakly conformal (i.e. $\phi^*h = \lambda g$, $\lambda \geq 0$) then ϕ is harmonic if and only if ϕ is a minimal branched immersion in the sense of Gulliver-Osserman-Royden [G-O-R].

iv. In general, if $\phi: M \to N$ is an isometric immersion ($\phi^*h = g$) then ϕ is harmonic if and only if ϕ is a minimal immersion.

v. If dim $M = 2$ and N is a symmetric space, harmonic maps are known to the physicists as *nonlinear σ-models*. Of particular interest in this context is the case where $N = \mathbb{C}P^n$ a complex projective space and $M = S^2$ the Riemann sphere (equivalently : finite action solutions on $M = \mathbb{R}^2$). Such σ-models are studied as a prototype non-abelian field theory being easier to handle than 4-dim Yang-Mills field while having similar qualitative properties: conformal invariance, existence of instantons and topological charges. For $N = \mathbb{C}P^1$ these σ-models occur in the study of ferromagnetism (see e.g. [B-P]).

vi. Finally, let us mention a result of Lichnerowicz to which we shall return below: if ϕ is a holomorphic map of almost Kähler manifolds then ϕ is harmonic [Li].

A fundamental question in the theory of harmonic maps is

"When can a given map be deformed into a harmonic map?"

i.e. what homotopy classes contain harmonic maps? In the case where N has non-positive sectional curvatures Eells and Sampson provide an affirmative answer for compact M and N by analytical procedures [E-S] and indeed variational approaches work well in this setting. But for many interesting cases, e.g. symmetric spaces of compact type, such methods do not work and so an alternative approach must be sought.

B. Let $U \subset R^2 = \mathbb{C}$ and consider maps $\phi : U \to \mathbb{C}^n$. Here the harmonic map equation is

$$\Delta \phi^i = \left(\frac{\partial^2}{\partial x^2} + \frac{\partial^2}{\partial y^2} \right) \phi^i = 0, \qquad 1 \le i \le n,$$

or, putting

$$\frac{\partial}{\partial z} = \frac{1}{2} \left(\frac{\partial}{\partial x} - i \frac{\partial}{\partial y} \right)$$

$$\frac{\partial}{\partial \bar{z}} = \frac{1}{2} \left(\frac{\partial}{\partial x} + i \frac{\partial}{\partial y} \right)$$

we have

$$\Delta \phi^i = \frac{1}{4} \frac{\partial^2}{\partial z \partial \bar{z}} \phi^i = 0.$$

Thus if $\frac{\partial}{\partial \bar{z}} \phi^i$ vanishes for each i, ϕ is harmonic. But $\frac{\partial}{\partial \bar{z}}$ is just the Cauchy-Riemann operator so that we have shown that holomorphic functions are harmonic. In general we make the following definitions:

Let (M,g) be an almost Hermitian manifold with Kähler form ω.

M is *cosymplectic* if ω is co-closed i.e. $d^*\omega = 0$

M is *(1,2)symplectic* if $(d\omega)^{1,2} = 0$.

Then we have

Theorem 1.1 [Li] Let $\phi : (M,g) \to (N,h)$ be an (anti-)holomorphic map of almost Hermitian manifolds where M is cosymplectic and N is (1,2) symplectic. Then ϕ is harmonic.

N.B. This extends the above since Kähler $\Rightarrow$ almost Kähler $\Rightarrow$ (1,2) symplectic $\Rightarrow$ cosymplectic.

Holomorphic maps are rather special harmonic maps for a number of reasons:

- if N is almost Kähler the (anti-)holomorphic maps are absolute minimizers of the energy in their homotopy classes [Li].

- they are solutions of a system of first-order elliptic differential equations. For dim $M = 2$, they are the "instanton" solutions for the nonlinear σ-models: the analogues of the (anti-)self-dual connections in the 4-dim Yang-Mills theory.

- if M and N are complex manifolds (i.e. the almost complex structures are integrable) then holomorphic maps are (at least in principle) more tractable than harmonic maps as the methods of complex Differential Geometry and Algebraic Geometry may be applied.

Thus it is reasonable to attempt to study harmonic maps by trying to reduce the problem to the study of holomorphic maps. This is the essence of the twistor approach to harmonic maps.

C. Essentially, the use of complex variable methods to study harmonic maps goes back to Weierstrass (c.f. [L], [H]) at least in the case of minimal surfaces in $\mathbb{R}^3$.

However, I choose to begin with the results of Calabi in 1967 [C]. Calabi established a 1-1 correspondence between full harmonic maps $S^2 \to S^{2m}$ and full "totally isotropic" holomorphic maps $S^2 \to \mathbb{C}P^{2m}$. (A map is full if its image is not contained in any proper totally geodesic subspace). The total isotropy condition involves (complex) orthogonality of iterated derivatives.

These ideas were taken up by a number of physicists and mathematicians who considered harmonic maps of a Riemann surface into a complex projective space ($\mathbb{C}P^n$ σ-models). Their construction is the following:

Let $f: M^2 \to \mathbb{C}P^n$ be holomorphic and define maps

$$W^i: M^2 \to G_{i+1, n+1} \text{ into Grassmannians by}$$
$$W^i = \text{span } (\tilde{f}, \partial\tilde{f}, \ldots \partial^{(i)}\tilde{f})$$

where $\partial = \frac{\partial}{\partial z}$, z a complex parameter on M^2 and $\tilde{f}$ a local representative of f in homogeneous coordinates. The W^i are in fact well-defined holomorphic maps into Grassmannians and if f is full we have $W_i \subseteq W_{i+1}$ for $i < N$.

Then Gram-Schmidt orthogonalisation of the resulting flag gives harmonic maps, that is, putting

$$f_{(i)} = W^\perp_{i-1} \cap W_i \quad , \quad i \leq n$$

defines a harmonic map $f_{(i)}: M^2 \to \mathbb{C}P^n$. The harmonic maps so obtained are 'complex isotropic', i.e. they satisfy an orthogonality condition on $\frac{\partial}{\partial z}$ and $\frac{\partial}{\partial \bar{z}}$ iterated derivatives.

Further, it can be shown that all such harmonic complex isotropic maps $M^2 \to \mathbb{C}P^n$ arise from holomorphic maps by the above procedures. Lastly, if M^2 is the Riemann sphere, it turns out that all harmonic maps $M^2 \to \mathbb{C}P^n$ are complex isotropic (this is a consequence of the fact that S^2 admits no holomorphic differentials). Thus we have the following Classification Theorem due to Din-Zakrewski, Glaser-Stora and Eells-Wood [DZ], [GS], [EW1], [EW2]:

Theorem 1.2 There is a 1-1 correspondence between full harmonic maps $\phi : S^2 \to \mathbb{C}P^n$ and pairs (f,r) where $f : S^2 \to \mathbb{C}P^n$ is a full holomorphic map and $0 \leq r \leq n$ given by

$$(f,r) \Rightarrow f_{(r)}.$$

Thus the study of harmonic maps $S^2 \to \mathbb{C}P^n$ reduces completely to the study of purely holomorphic objects.

These results have been extended to harmonic maps into Grassmannians by several authors see [Ra], [BW], [ErW], [CW] and will be considered in more detail in Chapter II.

D. In the setting of paragraph C the range manifold carries a natural complex structure (indeed Kähler structure) with respect to which we could consider holomoprhic maps. In most situations this will not be the case. We solve this problem by considering all possible almost complex structures on N simultaneously, that is, we introduce an auxiliary almost complex manifold fibring over N - the bundle of almost complex structures on N.

Let N be even-dimensional and orientable. Let $\pi : J(N) \to N$ be the fibre bundle over N whose fibre at x is the space of all almost complex structures on T_xN compatible with the metric i.e.

$$J_xN = \{J \in \text{End}(T_xN) : J^2 = -\text{Id}, \; J \text{ is skew}\}.$$

Abstractly, if $F(N)$ is the orthonormal frame bundle of N

$$J(N) = F(N) \times_{O(n)} O(n)/U(n).$$

Now $J(N)$ admits an almost complex structure conventionally denoted J_2, with the following interesting property:

Theorem 1.3 [ESa], [R] — Let M be almost Hermitian cosymplectic and $\psi : M \to J(N)$ holomorphic then $\pi_0\psi : M \to N$ is harmonic.

Remark A submersion of an almost complex manifold onto a Riemannian manifold is called a *twistor fibration* if it has the property of (1.3).

Of course, the question arises — do all harmonic maps $M \to N$ arise as projections of maps into $J(N)$? It turns out that for a satisfactory answer we must restrict attention to the case where M is a Riemann surface. Then the answer is affirmative in the following cases:

- if $\dim N = 4$. This is the theorem of Eells-Salamon [ESa],
- if N is Kähler [ESa],
- if $M = S^2$ and N is a compact inner Riemannian symmetric space.

It is a problem with this approach that the almost complex structure J_2 is *never* integrable. Thus the above results do not convert the harmonic map problem into one in the holomorphic category. However, as we shall see in the sequel, we can often use the geometry of the situation to find sub-bundles of $J(N)$ which carry an integrable complex structure that is related to J_2 in a simple way. In such situations we shall find that algebraic-geometric methods can be applied to construct harmonic maps.

Moreover, it is likely that this approach to the study of harmonic maps will provide the correct abstract framework in which to understand the explicit constructions of paragraph C.

Finally, we conclude by mentioning some applications of the twistor approach to the structure theory of harmonic maps:

- for harmonic maps $M^2 \to N^4$ there is a twistor degree (cf. [ESa]),

- for harmonic maps $S^2 \to G/H$, G/H an inner Riemannian symmetric space, there are 'Bäcklund transforms' producing new harmonic maps into symmetric spaces (cf [Bl]),

- there is an S^1-action on the space of harmonic maps $S^2 \to G$, G a compact Lie group ([U]).

E. We conclude this chapter with a short description of the contents of the following chapters.

In Chapter II we will discuss the explicit construction of harmonic maps into Grassmannians without reference to twistor spaces.

In Chapter III the bundle of almost complex structures will be described together with its J_2 complex structure.

In Chapter IV we will deal with the question of which harmonic maps arise as projections of holomorphic maps.

In Chapter V we shall examine some applications of the foregoing theory.

Supplementary References for Chapter 1

- for more information on harmonic maps, the Reader can do no better than refer to the survey articles of Eells-Lemaire [EL1], [EL2].

- for complex Differential Geometry, the books of Chern [C] and Wells [W] are useful while for Algebaric Geometry the best reference is Griffiths & Harris [GH].

II. Construction of 2-dim σ-models

A. Let $\mathbb{G}_{r,n}$ denote the complex Grassmannian of r-dimensional subspaces of $\mathbb{C}^n$. In particular $\mathbb{G}_{1,n} = \mathbb{C}P^{n-1}$ - (n-1)-dimensional complex projective space. U(n) acts on $\mathbb{G}_{r,n}$ in the obvious way giving it the structure of a Hermitian symmetric space $\frac{U(r)}{U(r) \times U(n-r)}$.

We define the tautological sub-bundle T of $\mathbb{G}_{r,n} \times \mathbb{C}^n$ by decreeing that the fibre of T at $W \in \mathbb{G}_{r,n}$ is W itself. Thus

$$T = \{(W,v) \in \mathbb{G}_{r,n} \times \mathbb{C}^n : v \in W\}.$$

Now there is a natural connection preserving isomorphism $h : T^{(1,0)}\mathbb{G}_{r,n} \to \text{Hom}(T,T^\perp)$ given by

$$h(X)\sigma = \pi_T^\perp(X \cdot \sigma)$$

where $X \in T^{(1,0)}\mathbb{G}_{r,n}$, σ is a local section of T and $\pi_T^\perp : \mathbb{G}_{r,n} \times \mathbb{C} \to T^\perp$ is orthogonal projection.

Henceforth we shall identify $T^{(1,0)}\mathbb{G}_{r,n}$ and $\text{Hom}(T,T^\perp)$ without further comment.

Now let M be a manifold. Then there is a 1-1 correspondence between maps $\phi : M \to \mathbb{G}_{r,n}$ and rank r sub-bundles of $M \times \mathbb{C}$ given by

$$\phi \rightsquigarrow \phi^{-1}T.$$

We shall denote $\phi^{-1}T$ by $\underline{\phi}$.

Now we have an isomorphism $\phi^{-1}T^{(1,0)}\mathbb{G}_{r,n} \cong \text{Hom}(\underline{\phi},\underline{\phi}^\perp)$ and if $\partial^{(1,0)}\phi$ denotes the (1,0) part of the differential of ϕ then we have

$$\partial^{(1,0)}\phi(X)\sigma = \pi_{\underline{\phi}}^\perp(X \cdot \sigma)$$

for $X \in TM^\mathbb{C}$, σ a local section of $\underline{\phi}$.

B. With these preliminaries out of the way let us turn to the construction of harmonic maps into $\mathbb{G}_{r,n}$. It is necessary to restrict attention to the case where M is a Riemann surface.

Let z be a complex parameter on M^2 then the harmonic map equation is

$$0 = \nabla d\phi\left(\frac{\partial}{\partial \bar{z}}, \frac{\partial}{\partial z}\right) = \phi^{-1}\nabla^{\mathbb{G}_{r,n}}_{\frac{\partial}{\partial \bar{z}}} \phi_*\left(\frac{\partial}{\partial z}\right) - \phi_*\left(\nabla^{M}_{\frac{\partial}{\partial \bar{z}}} \frac{\partial}{\partial z}\right)$$

$$= \phi^{-1}\nabla^{\mathbb{G}_{r,n}}_{\frac{\partial}{\partial \bar{z}}} \phi_*\left(\frac{\partial}{\partial z}\right) = \phi^{-1}\nabla^{\mathbb{G}_{r,n}}_{\frac{\partial}{\partial \bar{z}}} \left(\partial^{(1,0)}\phi\left(\frac{\partial}{\partial z}\right) + \partial^{(0,1)}\phi\left(\frac{\partial}{\partial z}\right)\right)$$

Now $\mathbb{G}_{r,n}$ is Kähler so $\phi^{-1}\nabla$ preserves the type decomposition of $\phi^{-1}TN^{\mathbb{C}}$ so that we conclude that ϕ is harmonic if and only if

$$\phi^{-1}\nabla^{\mathbb{G}_{r,n}}_{\frac{\partial}{\partial \bar{z}}} \partial^{(1,0)}\phi\left(\frac{\partial}{\partial z}\right) = 0 \tag{1}$$

or equivalently, since $\nabla d\phi$ is symmetric, if

$$\phi^{-1}\nabla^{\mathbb{G}_{r,n}}_{\frac{\partial}{\partial z}} \partial^{(1,0)}\phi\left(\frac{\partial}{\partial \bar{z}}\right) = 0 .$$

In our case, equation (1) has a simple geometrical interpretation. Results of Koszul-Malgrange [K-M] show that any complex bundle with connection ∇ over a Riemann surface has a natural holomorphic structure so that sections are holomorphic if and only if they are annihilated by $\nabla_{\frac{\partial}{\partial \bar{z}}}$. Now all sub-bundles of $M^2 \times \mathbb{C}^n$ inherit a connection and hence a holomorphic structure from the flat connection on $M^2 \times \mathbb{C}^n$. In particular $\underline{\phi}$ and $\underline{\phi}^{\perp}$ are holomorphic vector bundles in a natural way and it follows from (1) and the fact that the isomorphism h is connection-preserving that

Proposition 2.1

$\phi : M^2 \to \mathbb{G}_{r,n}$ is harmonic if and only if $dz \otimes \partial^{(1,0)}\phi\left(\frac{\partial}{\partial z}\right)$ is a holomorphic section of $T^*_{(1,0)}M \otimes \text{Hom}(\underline{\phi}, \underline{\phi}^\perp)$.

Now we can construct new harmonic maps as follows:

Set

$$s = \max_{z \in M} \text{rk } \partial^{(1,0)}\phi\left(\frac{\partial}{\partial z}\right) .$$

Then since $\partial^{(1,0)}\phi \cdot \frac{\partial}{\partial z}$ is holomorphic, $\dim \text{Im} \partial^{(1,0)}\phi(\frac{\partial}{\partial z}) = s$ off an isolated set of points and so defines a holomorphic vector subbundle of $\underline{\phi}$ off that isolated set. In fact, since $\dim M = 2$ we can extend the bundle over the isolated set to get a well-defined subbundle of $\underline{\phi}$ over M^2. We call this bundle the ∂'-*Gauss bundle of* ϕ, denoted $G'(\phi)$.

Thus

$$G'(\phi)_x = \text{Im } \partial^{(1,0)}_x \phi\left(\frac{\partial}{\partial z}\right) \quad \text{a.e } x.$$

Similarly we may define the ∂''-Gauss bundle of ϕ, $G''(\phi)$, with

$$G''(\phi)_x = \text{Im } \partial^{(1,0)}_x \phi\left(\frac{\partial}{\partial \bar{z}}\right) \quad \text{a.e } x.$$

Now we have

Theorem 2.2 If $\phi : M^2 \to \mathbb{G}_{r,n}$ is harmonic then $G'(\phi)$, $G''(\phi)$ define harmonic maps into $\mathbb{G}_{s,n}$, $\mathbb{G}_{t,n}$ where s (resp. t) are the maximal ranks of $\partial^{(1,0)}\phi(\frac{\partial}{\partial z})$ (resp. $\partial^{(1,0)}\phi(\frac{\partial}{\partial \bar{z}})$).

Remark In the notation of Ch.I (C), $G'(f) = \underline{f}_{(1)}$.

Thus given a harmonic map we may produce new ones essentially by differentiation. In particular, we may start with a holomorphic map and by repeated differentiation produce a series of harmonic maps. This is the basis of the classification theorem for $\mathbb{C}P^n$ σ-models.

C. Of course, we must identify which harmonic maps we obtain by differentiating holomorphic maps in this way. For this it is necessary to introduce the notion of 'complex isotropy':

Put $\underbrace{G'(G'\ldots(G'(\phi)))}_{k \text{ times}} = G^{(k)}(\phi)$ and define $G^{(-k)}(\phi)$

similarly for ∂''-Gauss bundles. Then ϕ is complex isotropic if

$$G^{(i)}(\phi) \perp G^{(j)}(\phi) \quad \forall\ i \neq j.$$

Concerning complex isotropy we note two facts:

Proposition 2.3

(a) if ϕ is complex isotropic so is $G^{(i)}(\phi)\ \forall\ i$,

(b) ϕ is complex isotropic if and only if

$$\text{Im}\left(\nabla_{\frac{\partial}{\partial \bar{z}}}\right)^{\alpha} \partial^{(1,0)} \phi\left(\frac{\partial}{\partial \bar{z}}\right) \perp \text{Im}\left(\nabla_{\frac{\partial}{\partial z}}\right)^{\beta} \partial^{(1,0)} \phi\left(\frac{\partial}{\partial z}\right) \quad \forall \alpha, \beta \geqq 0$$

where ∇ is the pull-back of the Levi-Civita connection on $T^{(1,0)}\mathbb{C}_{r,n}$. It is immediate from (b) that if ϕ is holomorphic then ϕ is complex isotropic (since $\partial^{(1,0)}\phi(\frac{\partial}{\partial z})$ would vanish identically) and so we may only hope to obtain isotropic harmonic maps by differentiating holomorphic ones. However, if our range is a complex projective space we do obtain all complex isotropic harmonic maps in this way.

To see this, let $\phi: M^2 \to \mathbb{C}P^{n-1}$ be harmonic and complex isotropic. Then since rank $\underline{\phi} = 1$ we have that rank $G^{(i)}(\phi) \leq 1$ for all i. Further, for $i \geq 0$ if $G^{(-i-1)}(\phi) = 0$ then $G^{(-i)}(\phi)$ is holomorphic since $G^{(-i-1)}\phi = \text{Im } \partial^{(0,1)}(G^{(-i)}(\phi))(\frac{\partial}{\partial z})$ a.e.

Now since all the Gauss bundles are mutually orthogonal, for $|i|$ large enough, the $G^{(i)}(\phi)$ must vanish as n non-zero Gauss bundles would exhaust $M^2 \times \mathbb{C}^n$. Thus if s is the least number for which $G^{(-s-1)}(\phi) = 0$ then $G^{(-s)}(\phi)$ is holomorphic and it is easy to show that $\underline{\phi} = G^{(s)}(G^{(-s)}(\phi))$.

In the case that $M^2 = S^2$, the Riemann sphere, we can say more: we define differentials by

$$\eta_\alpha = \text{Trace} \left(\partial^{(1,0)} \phi\left(\frac{\partial}{\partial z}\right) \right)^* \nabla^\alpha_{\frac{\partial}{\partial z}} \partial^{(1,0)} \phi\left(\frac{\partial}{\partial z}\right) dz^{\alpha+2}$$

and we find that if $\eta_{\alpha-1}$ vanishes then η_α is holomorphic so that since S^2 admits no holomorphic differentials, each η_α vanishes. Now since rk $\underline{\phi} = 1$ for maps into $\mathbb{C}P^{n-1}$ it is easy to conclude from the above and Prop. 2.3(b) that

Proposition 2.4

All harmonic maps $S^2 \to \mathbb{C}P^{n-1}$ are complex isotropic.

Thus adding a fullness ingredient to guarantee uniqueness we recover the classification theorem 1.2 of Chapter I.

D. For maps $\phi: M^2 \to \mathbb{G}_{r,n}$, $r > 1$, the theory is much less straightforward for a number of reasons:

— not all complex isotropic harmonic maps arise by differentiation of holomorphic maps.

— worse, even for $M^2 = S^2$ not all harmonic maps are complex isotropic. In fact, there is no information available on the higher derivatives of ϕ, essentially because $G_{r,n}$ only has constant holomorphic sectional curvature if $r = 1$.

However, the construction problem for harmonic maps $S^2 \to G_{r,n}$, $r > 1$, has been studied by many authors, e.g. Ramanathan ($G_{2,4}$), Aithal ($G_{2,5}$), Chern-Wolfson ($G_{2,n}$), Burstall-Wood ($G_{r,n}$, $r \leq 5$) and Uhlenbeck ($G_{r,n}$ all r,n) ([Ra], [A], [C-W], [B-W], [U]).

What is required in these cases is a new way of building new harmonic maps from old. Let us describe one such procedure:

Let $\phi : M^2 \to G_{r,n}$ be a harmonic map and let $\underline{\alpha} \subset \underline{\phi}$, $\underline{\beta} \subset \underline{\phi}^\perp$ be holomorphic subbundles such that

$$\partial^{(1,0)} \phi \left(\frac{\partial}{\partial z} \right) \underline{\alpha} \subset \underline{\beta}$$

$$\left(\partial^{(1,0)} \phi \left(\frac{\partial}{\partial z} \right) \right)^* \underline{\beta} \subset \underline{\alpha} \quad .$$

Then putting $\underline{\tilde{\phi}} = \underline{\alpha}^\perp \cap \underline{\phi} + \underline{\beta}$ defines a new harmonic map $\tilde{\phi} : M^2 \to G_{s,n}$.

In the terminology of Burstall-Wood this procedure is called 'forward replacement of $\underline{\alpha}$ by $\underline{\beta}$', in that of Uhlenbeck it is called 'adding a uniton'.

Example Taking $\underline{\alpha} = \underline{\phi}$, $\underline{\beta} = G'(\phi)$ we recover Theorem 2.2.

Now judicious choice of $\underline{\alpha}$ and $\underline{\beta}$ yield the following theorem constructing all harmonic maps $S^2 \to G_{r,n}$.

Theorem 2.5 Let $\phi : S^2 \to G_{r,n}$ be harmonic. Then there is a sequence of harmonic maps $\phi_0, \ldots, \phi_N$ of S^2 into Grassmannians such that

(i) ϕ_0 is holomorphic

(ii) $\phi_N = \phi$

(iii) ϕ_{i+1} is obtained from ϕ_i by forward replacement of a holomorphic subbundle of $\underline{\phi}_i$.

Remark This theorem was proved independently and using quite different viewpoints by Chern-Wolfson in case of $\mathbb{G}_{2,n}$, Burstall-Wood in case of $\mathbb{G}_{r,n}$, $r \leq 5$, and by Uhlenbeck in complete generality. Indeed Uhlenbeck's method treats all harmonic maps of S^2 into the unitary group.

It should be pointed out that while we can now *construct* all harmonic maps $S^2 \to \mathbb{G}_{r,n}$ we still have no *classification* of such maps as we do for maps $S^2 \to \mathbb{CP}^{n-1}$. The construction procedures of Theorem 2.5 are highly non-unique and it seems that a new approach is needed to provide a classification in terms of holomorphic data.

Notes and References for Chapter II

- for information on the classification of harmonic maps $S^2 \to \mathbb{CP}^n$ refer to [EW2]

- the viewpoint of this chapter is that of Burstall-Wood [BW]; for a moving frames approach see Chern-Wolfson [CW].

III. Twistor spaces

A. Let (N,h) be an orientable even-dimensional n-manifold with metric h. Let $J(N)$ denote the fibre bundle over N whose fibre at $x \in N$ is the space of almost complex structures on $T_x N$ compatible with the metric i.e.

$$J_x(N) = \{J \in o(T_x N) : J^2 = -\text{Id}\}.$$

(For any inner product space V, $o(V)$ denotes the skew-symmetric endomorphisms on V).

To understand the geometry of this bundle let us begin by examining the geometry of the fibre:

Denote by V an even-dimensional Euclidean space and let $J(V)$ be the space of Hermitian almost complex structure on V. Then $O(V)$ acts transitively on $J(V)$ by conjugation so that $J(V) \subset o(V)$ is an adjoint orbit of $O(V)$ in its Lie algebra. As such it admits an invariant complex structure which is given as follows:

Fix $J_o \in J(V)$, then the stabiliser of J_o, H_{J_o}, has Lie algebra Ker ad J_o for which Im ad J_o is an H_{J_o} invariant complement. Thus

$$T_{J_o} J(V) \cong \text{Im ad } J_o.$$

Further it is easily checked that $\frac{1}{2}$ ad J_o defines an H_{J_o} invariant almost complex structure on Im ad J_o so that $J(V)$ acquires an $O(V)$ invariant almost complex structure given at J_o by $\frac{1}{2}$ ad J_o. In fact, this complex structure is integrable.

Now let us turn to $J(N)$, which is associated to the orthogonal frame bundle $F(N)$ of N. In fact

$$J(N) = F(N) \times_{O(n)} J(\mathbb{R}^n).$$

The Levi-Civita connection induces a horizontal distribution on $F(N)$

and hence one on $J(N)$ so that we have a splitting

$$TJ(N) = V \oplus H$$

where V is the vertical distribution with respect to the projection $\pi : J(N) \to N$ and H is isomorphic to $\pi^{-1}TN$.

Now $\pi^{-1}TN$ carries a tautological almost complex structure given by

$$J^H_J = J,$$

identifying $(\pi^{-1}TN)_J$ with $T_{\pi(J)}N$, while V inherits an almost complex structure, J^V, from the $O(n)$-invariant one on $J(R^n)$.

Thus we define two almost complex structures on $J(N)$ by

$$J_1 = \begin{cases} J^H & \text{on } H = \pi^{-1}TN \\ J^V & \text{on } V \end{cases}$$

$$J_2 = \begin{cases} J^H & \text{on } H \\ -J^V & \text{on } V \end{cases}$$

These two complex structures which differ only by a sign on the vertical distribution have very different properties, for instance, J_1 is sometimes integrable (under stringent conditions on the curvature of N) while J_2 is *never* integrable ([S]). The J_1 structure is unchanged under conformal change of metric (and hence Levi-Civita connection) on N unlike J_2.

The fibration $(J(N), J_1) \to N$ is a natural generalisation of the Penrose fibration $\mathbb{C}P^3 \to S^4$ studied by Atiyah et al. (cf [At]) but it is the less tractable J_2 structure which will be of the most importance in the sequel.

B. Now let M be a manifold and $\psi : M \to J(N)$ be a smooth map. Let $\phi : M \to N$ be given by $\phi = \pi_0 \psi$. Then we have

$$\phi^{-1}TN = \psi^{-1}(\pi^{-1}TN) = \psi^{-1}H.$$

Thus $\phi^{-1}TN$ inherits an almost complex structure from J^H on H. In fact, at $x \in M$, the almost complex structure on $(\phi^{-1}TN)_x = T_{\phi(x)}N$ is precisely $\psi(x) \in J(N)$.

Conversely, given $\phi : M \to N$, an almost complex structure on $\phi^{-1}TN$ determines a map into $J(N)$ covering ϕ.

When is ψ holomorphic with respect to J_1 or J_2? From Rawnsley [R] we have

Theorem 3.1   Let $\psi : M \to J(N)$ be a smooth map where M is an almost complex manifold. Let $\phi = \pi_0 \psi$ and

$$\phi^{-1}TN^{\mathbb{C}} = \underline{\psi}^+ \oplus \underline{\psi}^-$$

be the induced splitting of $\phi^{-1}TN^{\mathbb{C}}$ into $+i$ and $-i$ eigenbundles of the almost complex structure given by ψ.

Then (i) ψ is J_1 holomorphic if and only if

(a) $\phi_*(T^{(1,0)}M) \subset \underline{\psi}^+$

(b) $(\psi^{-1}\nabla)_Z C^\infty(\underline{\psi}^+) \subset C^\infty(\underline{\psi}^+)$ $Z \in T^{(1,0)}M$

(ii) ψ is J_2 holomorphic if and only if

(a) $\phi_*(T^{(1,0)}M) \subset \underline{\psi}^+$

(b) $(\phi^{-1}\nabla)_{\overline{Z}} C^\infty(\underline{\psi}^+) \subset C^\infty(\underline{\psi}^+)$ $Z \in T^{(1,0)}M$.

Remarks   (i) Condition (a) corresponds to holomorphicity of the horizontal component of ψ and (b) to holomorphicity of the vertical component.

(ii) The vertical condition for J_2 holomorphicity is equivalent to the demand that $\underline{\psi}^+$ be a holomorphic subbundle of $\phi^{-1}TN^{\mathbb{C}}$ in the holomorphic structure induced by $\phi^{-1}\nabla$ (which is, of course, the pull-back of the Levi-Civita on N).

(iii) If ψ is simultaneously holomorphic with respect to J_1 and J_2 then $\psi_*(TM) \subset H$ and we say that ψ is *horizontal holomorphic*. This is equivalent to condition (a) and $\underline{\psi}^+$ being a parallel subbundle of $\phi^{-1}TN^{\mathbb{C}}$.

We are now in a position to understand the significance of the J_2 structure.

Theorem 3.2 Let (M,J) be almost Hermitian cosymplectic and $\psi : M \to J(N)$ be holomorphic with respect to J_2. Then $\pi_0 \psi = \phi$ is harmonic.

Proof Let $E_1, JE_1, \ldots, E_m, JE_m$ be an orthonormal basis for TM. Then putting $Z_i = E_i - iJE_i$, $Z_1, \ldots, Z_m$ are a local frame for $T^{(1,0)}M$ and the harmonic map equation is

$$\tau_\phi = \text{Trace } \nabla d\phi = \nabla d\phi(Z_i, \overline{Z}_i) = 0 .$$

Now

$$\nabla d\phi(Z_i, \overline{Z}_i) = (\phi^{-1}\nabla^N)_{\overline{Z}_i} \phi_*(Z_i) - \phi_*(\nabla^M_{\overline{Z}_i} Z_i) .$$

The cosymplectic condition on M means that

$$\nabla^M_{\overline{Z}_i} Z_i \in T^{(1,0)}M$$

so by condition (a) $\phi_*(\nabla_{\overline{Z}_i} Z_i) \subset \underline{\psi}^+$.

Further $(\phi^{-1}\nabla^N)_{\overline{Z}_i} \phi_*(Z_i) \subset \underline{\psi}^+$ by (a) and (ii)(b).

Thus $\tau_\phi \in \underline{\psi}^+$. But τ_ϕ is real so

$$\tau_\phi \subset \underline{\psi}^+ \cap \overline{\underline{\psi}^+} = \underline{\psi}^+ \cap \underline{\psi}^- = \{0\} \text{ whence } \phi \text{ is harmonic.}$$

Definition   Let N be a Riemannian manifold. A *twistor fibration* is a Riemannian submersion of an almost Hermitian manifold Z (*the twistor space*) onto N with the following property:

If M is almost Hermitian cosymplectic and $\psi : M \to Z$ is holomorphic then $\pi_o \psi : M \to N$ is harmonic.

Thus Theorem 3.2 shows that $\pi : (J(N), J_2) \to N$ is a twistor fibration. In fact, most twistor fibrations we shall consider will arise as submanifolds of J(N).

C. For many applications one would like twistor fibrations with more tractable properties than those of J(N) e.g. a (1,2)-symplectic metric, an integrable or even Kähler J_1 and so on. Thus one is led to consider smaller twistor spaces. One method of obtaining such spaces is due to Rawnsley [R]:

Let Z be a fibre bundle over N with each fibre a complex manifold with complex structure varying smoothly from fibre to fibre. Suppose also that Z has a horizontal distribution and there is a fibre-preserving map

$$f : Z \to J(N)$$

such that

   (i)   f is holomorphic on fibres

   (ii) f preserves the horizontal distributions.

Then the horizontal distribution on Z acquires an almost complex structure by pulling back that on J(N) by f and we may define J_1 and J_2 as before and f will then be holomorphic with respect to both the J_1 structures and the J_2 structures. In particular (Z, J_2) will give rise to a twistor fibration.

The twistor spaces arising from the above constructions are called 'generalised twistor spaces' by Rawnsley.

Examples (i) Fix an orientation on N and let $J^+(N)$ denote the bundle of almost Hermitian structures on N compatible with the orientation. $J^+(N)$ is an SO(2n) fibre bundle with fibre SO(2n)/U(n) which is a Hermitian symmetric space. $J^+(N)$ inherits a horizontal distribution from the Levi-Civita connection and the inclusion $J^+(N) \to J(N)$ makes $J^+(N)$ into a generalised twistor space. This is the twistor space studied by Eells and Salamon [ESa], [S].

(ii) Let (N,J) be a Kähler manifold and let $Z = G_r(T^{(1,0)}N)$ be the Grassmannian bundle of r-planes in $T^{(1,0)}N$. If F(N) denotes the U(n) bundle of unitary frames on N then

$$Z = F(N) \times_{U(n)} \mathbb{G}_{r,n}.$$

Z inherits a horizontal distribution from that on F(N) given by the Levi-Civita connection (which is a U(n) connection since N is Kähler) and the fibres of Z acquire complex structures from the U(n)-invariant one on $\mathbb{G}_{r,n}$. Now define $f : Z \to J(N)$ by

$$f(W) = J \text{ on } (W + \overline{W}) \cap TN$$

$$= -J \text{ on } ((W + \overline{W}) \cap TN)^\perp$$

then f satisfies (i) and (ii) above and $G_r(T^{(1,0)}N)$ is a generalised twistor space.

Remark If $\dim_{\mathbb{C}} N \geq 3$ J_1 is integrable on $G_r(T^{(1,0)}N)$ if and only if the Bochner curvature tensor of N vanishes identically cf. [O-R].

(iii) Let G/K be a Riemannian symmetric space of compact type and let the associated symmetric decomposition of Lie algebras be

$$G = k + p, \quad [k,k] \subset k, \quad [k,p] \subset p, \quad [p,p] \subset k.$$

Suppose there exists an element $\xi \in k$ whose centraliser is contained in K (for this it is necessary and sufficient that G/K be *inner* i.e. rankG = rankK). Denote the stabiliser of ξ by H.

With respect to any G-invariant metric on G, adξ is skewsymmetric and so has imaginery eigenvalues which come in conjugate-pairs.

So for $\lambda \in \mathbb{R}$ let

$$k^\lambda = \{v \in k^{\mathbb{C}} : [\xi, v] = i\lambda v\}$$

$$p^\lambda = \{v \in p^{\mathbb{C}} : [\xi, v] = i\lambda v\}$$

and we have

$$p^{-\lambda} = \overline{p^\lambda} \qquad k^{-\lambda} = \overline{k^\lambda}$$

$$[p^\lambda, p^\mu] \subset k^{\lambda+\mu}, [p^\lambda, k^\mu] \subset p^{\lambda+\mu}, [k^\lambda, k^\mu] \subset k^{\lambda+\mu}$$

$$G^{\mathbb{C}} = \sum_\lambda k^\lambda + p^\lambda$$

$k^0 = h^{\mathbb{C}}$ — the Lie algebra of H, $p^0 = \{0\}$.

Lastly, put

$$h^+ = \sum_{\lambda > 0} h^\lambda, \quad p^+ = \sum_{\lambda > 0} p^\lambda \qquad \text{and}$$

$$h^- = \overline{h^+}, \quad p^- = \overline{p^+}.$$

I claim that the homogeneous fibration G/H → G/K gives G/H the structure of a generalised twistor space. We can see this as follows:

Equip K/H with the K-invariant complex structure with +i eigenspace at the identity case given by k^+. (This is clearly integrable since $[k^+,k^+] \subset k^+$). Define a map

$$j : K/H \to J(p) \text{ by}$$

$$j(kH) = +i \text{ on } \mathrm{Adk}(p^+)$$
$$ -i \text{ on } \mathrm{Adk}(p^-).$$

Then j is clearly K-equivariant and can be shown to be holomorphic (this is a consequence of $[k^+, p^+] \subset p^+$).

Now $G \to G/K$ is a principal K-bundle and TG/K is associated to G via the adjoint representation on p. Thus j induces a map

$$f : G/H = G \times_K K/H \to J(G/K)$$

which is holomorphic on fibres (with respect to the left translate of the k^+ complex structure) and preserves horizontal distributions (where that on G/H is given by the left translates of p).

That f is holomorphic on fibres is an immediate consequence of the holomorphicity of j. That f preserves horizontal distributions follows since that on G/H comes from the canonical connection on G which induces the Levi-Civita connection on G/K since G/K is symmetric.

In fact, the complex structures J_1, J_2 on G/H are easy to identify: they are both G-invariant and at the identity coset have +i-eigenspaces given by

$$J_1 : k^+ + p^+$$
$$J_2 : k^- + p^+.$$

These complex structures are rather better behaved than their counterparts on $J(G/K)$:

(i) J_1 is always integrable and indeed Kählerian

(ii) J_2 is sometimes $(1,2)$-symmetric (this is so if $[k^+, p^+] = 0$).

Twistor spaces of this kind satisfying (ii) have been studied by Bryant [Br3] and Salamon [S].

The case where $(G/H, J_1)$ is a complex flag manifold (i.e. of the form $G^{\mathbb{C}}/P$ where $G^{\mathbb{C}}$ is a complex Lie group and P is a parabolic subgroup) has been studied by Burstall and Rawnsley. We shall return to this case in Chapter IV.

Remark The viewpoint in this chapter is that of Rawnsley [R] to which paper the reader is referred for many extensions, e.g. to spaces of f-structures.

IV. Twistor Lifts

In Chapter III we saw how to obtain harmonic maps as projections of holomorphic maps into twistor spaces. It is natural to enquire into when all harmonic maps arise in this way. For the most part it will be necessary to restrict attention to 2-dimensional orientable domains.

A. The first result obtained in this direction was that of Eells and Salamon concerning branched minimal surfaces in 4-manifolds:

Let (N,h) be an essential Riemannian 4-manifold and consider the bundle $\pi : J^+(N) \to N$ of Hermitian almost complex structures on N compatible with the orientation, i.e.

$$J^+(N) = \{J \in J(N) \text{ s.t. } *1(X_1 \wedge JX_1 \wedge X_2 \wedge JX_2) \geq 0 \text{ for any vectors } X_1, X_2\}.$$

Now let M^2 be a Riemann surface with local isothermal co-ordinate z and let $\phi : M^2 \to N$ be a non-constant conformal harmonic map. In this case the harmonic map equation reduces to

$$\left(\phi^{-1}\nabla^N_{\frac{\partial}{\partial \bar{z}}}\right) \phi_*\left(\frac{\partial}{\partial z}\right) = 0 \tag{1}$$

while the conformality condition is

$$\phi^* h^{(2,0)} \equiv 0 \quad \text{or} \quad h(\phi_*(\frac{\partial}{\partial z}), \phi_*(\frac{\partial}{\partial z})) = 0. \tag{2}$$

From (1) we see that $\phi_*(\frac{\partial}{\partial z})$ is a local holomorphic section of $\phi^{-1}TN^{\mathbb{C}}$ with respect to the complex structure induced by the Levi-Civita connection on N. Thus $\phi_*(\frac{\partial}{\partial z})$ only vanishes at isolated points and defines a holomorphic line subbundle of $\phi^{-1}TN^{\mathbb{C}}$ which can be extended across the zeros of $\phi_*(\frac{\partial}{\partial z})$.

Thus we have a line bundle L such that

(i) $\phi_*(\frac{\partial}{\partial z}) \subset L$

(ii) $\phi^{-1}\nabla^N_{\frac{\partial}{\partial \bar{z}}} C^\infty(L) \subset C^\infty(L)$ (L is holomorphic)

(iii) $h(C^\infty(L), C^\infty(L)) \equiv 0$ i.e. L is isotropic, this follows from (2).

Now $(L + \bar{L})$ is stable under conjugation and has complex dimension 2 and since TN and hence $\phi^{-1}TN$ is orientable we see that there is an isotropic line subbundle L_1 of $(L + \bar{L})$ unique up to conjugation.

Now $L \oplus L_1$ is a maximally isotropic subbundle of $\phi^{-1}TN^{\mathbb{C}}$ and so is the (1,0) space for a Hermitian almost complex structure on $\phi^{-1}TN^{\mathbb{C}}$. Further, we may assume that this almost complex structure is compatible with the orientation (otherwise replace L_1 with $\bar{L}_1$). Thus there is a map $\psi : M^2 \to J^+(N)$ for which, in the notation of Chapter III:

$$\underline{\psi}^+ = L \oplus L_1.$$

I claim that ψ is J_2-holomorphic.

Firstly we have

$$\phi_*(\frac{\partial}{\partial z}) \subset \underline{\psi}^+ .$$

Thus, it suffices to show that

$$\left(\phi^{-1}\nabla^N_{\frac{\partial}{\partial \bar{z}}}\right) C^\infty(\underline{\psi}^+) \subset C^\infty(\underline{\psi}^+) .$$

From (ii) we need only consider $C^\infty(L_1)$ so let $\sigma \in C^\infty(L_1)$. Then

$$h\left(\nabla_{\frac{\partial}{\partial \bar{z}}} \sigma, \sigma\right) = 0 \text{ since } L_1 \text{ is isotropic,}$$

$$h\left(\nabla_{\frac{\partial}{\partial \bar{z}}} \sigma, \phi_*(\tfrac{\partial}{\partial z})\right) = -h\left(\sigma, \nabla_{\frac{\partial}{\partial \bar{z}}} \phi_*(\tfrac{\partial}{\partial z})\right) = 0 \text{ since } \phi \text{ is harmonic}$$

so that

$$h\left(\nabla_{\frac{\partial}{\partial \bar{z}}} \sigma, C^\infty(\underline{\psi}^+)\right) = 0 \text{ whence}$$

$$\nabla_{\frac{\partial}{\partial \bar{z}}} C^\infty(L_1) \subset C^\infty(\underline{\psi}^+) \text{ by the maximal isotropy of } \underline{\psi}^+.$$

The claim now follows from Theorem 3.1 and we have proved

Theorem 4.1 ([ESa])   There is a 1:1 correspondence between non-constant conformal harmonic maps $\phi : M^2 \to N^4$ and non-vertical J_2-holomorphic maps $\psi : M^2 \to J^+(N^4)$ given by the above construction.

Remarks   (i)   The uniqueness of the correspondence comes from the unique choice of L_1 compatible with the orientation.

(ii)   The map ψ constructed from ϕ above is essentially the Gauss map of ϕ.

(iii)   The assumption of 2-dimensional orientable domain was required for the following reasons:

(a)   To extend the definition of L across the zeros of $\phi_*(\tfrac{\partial}{\partial z})$ which requires the integrability of the complex structure induced by $\phi^{-1}\nabla^N$ and some single complex variable methods.

(b)   To ensure the stability of $C^\infty(L)$ under $\phi^{-1}\nabla^{N(0,1)}$ which in general would require the vanishing of $\nabla d\phi^{(1,1)}$ which for $\dim_{\mathbb{C}} M > 1$ is a stronger condition than harmonicity.

B. For $\dim N > 4$ it is not clear whether every harmonic $\phi : M \to N$ is covered by a J_2 holomorphic map although Rawnsley and Salamon have shown that such lifts exist locally (globally if $M = S^2$ or N is orientable), [S], [R].

However, when N has more structure, lifts can be built. For example, let N be a Kähler manifold with twistor space $G_1(T^{(1,0)}N) = \mathbb{P}(T^{(1,0)}N)$ discussed in Chapter III.

Let $\phi : M^2 \to N$ be a conformal non-anti-holomorphic harmonic map. Then if $\partial^{(1,0)}\phi(\frac{\partial}{\partial z})$ is non-zero at a given point it defines a complex line in $T^{(1,0)}N$ and hence a point in $G_1(T^{1,0}N)$. Using the integrability of the holomorphic structure on $\phi^{-1}T^{(1,0)}N$ induced by the Levi-Civita connection on N we can extend this map into $G_1(T^{(1,0)}N)$ across the zeros of $\partial^{(1,0)}\phi(\frac{\partial}{\partial z})$ and so we have a map $\psi : M^2 \to G_1(T^{(1,0)}N)$ covering ϕ with

$$\text{span } \partial^{(1,0)}\phi(\frac{\partial}{\partial z}) = \psi .$$

Similar arguments to those in paragraph A show that ψ is J^2 holomorphic and establish

Theorem 4.2 [E-Sa], [R]. There is a 1:1 correspondence between conformal harmonic non-anti-holomorphic maps $\phi : M^2 \to N$, N a Kähler manifold and J_2 holomorphic non-vertical maps $\psi : M^2 \to G_1(T^{(1,0)}N)$ given by

$$\phi = \pi_0 \psi , \qquad \psi = \text{span } \partial^{(1,0)}\phi(\frac{\partial}{\partial z}) \quad \text{a.e.}$$

C. The results considered above provide a *parametrisation* of certain classes of harmonic maps in terms of holomorphic maps into a twistor space. In general, such 1:1 correspondences are too much to hope for, but we may still try to construct a holomorphic map into a twistor space corresponding to a given harmonic map. Again, one may wish to find twistor lifts into better behaved, 'smaller' twistor spaces than $J(N)$.

One situation where such a program may be carried out is the case when N is a compact inner Riemannian symmetric space and M is the Riemann sphere S^2.

Now let $H \subset K$ be the centraliser of a torus and consider the homogeneous fibration $\pi : G/H \to G/K$. We may equip G/H with an integrable complex structure J_1 in the following way:

There is at least one parabolic subgroup P, of $G^{\mathbb{C}}$ such that $P \cap G = H$. Further, G acts transitively on $G^{\mathbb{C}}/P$ so that there is a natural diffeomorphism

$$G/H \cong G^{\mathbb{C}}/P$$

and G/H inherits a complex structure from that on $G^{\mathbb{C}}/P$. This complex structure is integrable and Kählerian. We now define an almost complex J_2 on G/H by reversing the orientation of J_1 on Ker π_*. Then G/H is a generalised twistor space as in Chapter III paragraph C and in particular $\pi : (G/H, J_2) \to G/H$ is a twistor fibration.

We now have

Theorem 4.3 Let $\phi : S^2 \to G/K$ be a harmonic map of the Riemann sphere into a compact inner Riemannian symmetric space. Then there is a complex flag manifold G/H, $H \subset K$ with a generalised twistor space structure as above and a J_2-holomorphic map $\psi : S^2 \to G/H$ such that

$\pi_0 \psi = \phi$. Let us briefly consider the main ingredients of the proof.

If $G = k + p$ is the symmetric decomposition of G/K, then $TG/K = G \times_K p \subset G/K \times g$ so that we may consider TG/K as a subbundle of the trivial bundle $G/K \times g$. (This identification is essentially by the moment map of TG/K). Thus we consider $d\phi$ as a g-valued 1-form on S^2 and we may equip the trivial bundle $S^2 \times g^{\mathbb{C}}$ with the g-connection $\nabla^\phi = d - d\phi$. This connection coincides with the pull-back of the Levi-Civita connection on $\phi^{-1} TG/K$ from which it follows that ϕ is harmonic if and only if $\phi_* \frac{\partial}{\partial z}$ is a local holomorphic section of $S^2 \times g^{\mathbb{C}}$ with the holomorphic structure induced by ∇^ϕ.

Now using the Birkhoff-Grothendieck classification theorem for holomorphic vector bundles on the Riemann sphere (see [G]) we obtain a holomorphic reduction of $M^2 \times g^{\mathbb{C}}$ as a $G^{\mathbb{C}}$ bundle to a parabolic subgroup P and hence a map $\psi : S^2 \to G/H = G/P$ with $H \subset K$ which is J_2-holomorphic in the vertical directions. Further $\underline{\psi}^+$ contains the images of all meromorphic sections of $\phi^{-1} TG/K$ with positive degree divisors. Now since $T^{(1,0)} S^2$ is spanned by a meromorphic section with degree 2 divisor and $d\phi | T^{(1,0)} S^2$ is holomorphic it follows that

$$\phi_*(T^{(1,0)} S^2) \subset \underline{\psi}^+$$

so that ψ is J_2 holomorphic in the horizontal directions as well.

Remarks (i) This result differs from Theorems 4.1 and 4.2 in that the proof is non-constructive; in most cases we cannot describe explicitly the holomorphic map ψ. However, in case $G = SU(n)$ the lift may be built explicitly from ϕ and its derivatives and the use of the

Birkhoff-Grothendieck theorem may be avoided, see [B2].

(ii) The above result may be extended to surfaces of higher genus if G/K is Hermitian symmetric and $d\phi$ has sufficiently many zeros relative to the genus. Here the main point is to replace the Birkhoff-Grothendieck theorem (valid only for S^2) with the Harder-Narasimhan decomposition of a holomorphic vector bundle into a flag of subbundles with semi-stable quotients. A good reference for the Harder-Narasimhan story is Atiyah-Bott [A-B].

D. Let us conclude this chapter by considering a result of Uhlenbeck which provides twistor lifts in a completely different way.

Let G be a compact Lie group with Maurer-Cartan form $\omega = g^{-1}dg$ and let $\phi : S^2 \to G$ be a harmonic map. In this case the harmonic map equation is

$$d^*(\phi^*\omega) = 0 . \tag{1}$$

We also have the pull-back of the Maurer-Cartan equations

$$d(\phi^*\omega)(X,Y) + [\phi^*\omega(X), \phi^*\omega(Y)] = 0 \quad X,Y \in TS^2 . \tag{2}$$

Denote $\phi^*\omega$ by $A = A_z dz + A_{\bar{z}} d\bar{z}$. Then we may interpret (2) as the assertion that the G-connection

$$d + A$$

on $S^2 \times G$ has zero curvature and (1) as the assertion that ϕ is a Hodge (Lorentz) gauge for the trivial connection.

Now define, for $\lambda \in \mathbb{C}\setminus\{0\}$, new connections

$$\nabla^\lambda = (\bar{\partial} + \frac{(1-\lambda)}{2} A_{\bar{z}}, \partial + \frac{(1-\lambda^{-1})}{2} A_z).$$

Then equations (1) and (2) are equivalent to the vanishing of the curvature of ∇^λ for all $\lambda \in S^1$. Since S^2 is simply connected it follows that each ∇^λ is gauge-equivalent to the trivial connection and so we have a smooth family of maps $\phi_\lambda : S^2 \to G$ with

$$\phi_{-1} = \phi \qquad \phi_1 = e, \quad \text{the identity in } G$$

$$\nabla^\lambda = d + \phi_\lambda^* \omega. \tag{3}$$

We interpret this result in the following way:

Let $\Omega G = L_1^2(S^1, 1; G, e)$ be the group of based loops in G of Sobolev class L_1^2. ΩG is a smooth infinite dimensional manifold and in fact a Banach Lie group. Further ΩG admits a left invariant complex structure which is Kählerian (see, for example, [P]).

Now let $\pi_{-1} : \Omega G \to G$ be given by evaluation at -1. Then given $\phi : S^2 \to G$ harmonic the above constructions produce a map

$$\Phi : S^2 \to \Omega G \text{ given by } \Phi(x)(\lambda) = \phi_\lambda(x) \text{ with } \pi_{-1} \circ \Phi = \phi,$$

and equation (3) can be interpreted as the assertion that Φ is holomorphic and satisfies a kind of strong horizontality condition. Thus harmonic maps into G from S^2 are covered by holomorphic maps into an *infinite-dimensional complex manifold*.

In fact, we may fit these results completely into our present context by means of the following theorem of Burstall:

Theorem 4.4 There exists a non-integrable left invariant almost complex structure J_2 on ΩG with respect to which $\pi_{-1} : \Omega G \to G$ is a twistor fibration.

J_2 coincides with the standard complex structure on the subspace of $T\Omega G$ where the maps Φ constructed above take their images so that Uhlenbeck's results provide a twistor lift for the above twistor fibration.

Remarks (i) For $G = U(n)$, Uhlenbeck's result is the starting point for her construction of harmonic maps of S^2 into G from holomorphic maps.

(ii) ΩG contains the complex flag manifolds G/H as conjugacy classes of homomorphisms $S^1 \to G$. These G/H are embedded totally geodesically and J_2-holomorphically for the J_2 structures on G/H considered in Chapter III; further $\pi_{-1}|G/H$ takes images in a copy of a symmetric space G/K contained in G (G/K is inner!) in a canonical way. Thus ΩG is in some sense a universal twistor space for symmetric spaces G/K.

Remarks and Supplementary References

- The viewpoint in paragraphs A and B is that of Rawnsley [R].
- The results of paragraph C are due to Burstall and Rawnsley. In this context [B1], [B2] are relevant.
- The results of Uhlenbeck discussed in paragraph D are contained in [U].

V. Applications

A. *Existence of harmonic maps*

We have seen in the preceding chapters how holomorphic curves in twistor spaces give rise to harmonic maps. However, the existence theory for holomorphic curves into almost complex manifolds is in fairly poor shape (see, however, the recent work of Gromov [Gr]) so that we do not necessarily gain information about the existence of harmonic maps.

One way round this difficulty is to consider horizontal holomorphic curves i.e. maps which are holomorphic with respect to both J_1 and J_2. If the twistor space has an integrable J_1 with respect to which the (1,0) horizontal distribution is holomorphic then, to find horizontal holomorphic curves one is reduced to solving a holomorphic differential equation. This approach has been used by Bryant, [Br1], [Br2] to prove existence theorems for harmonic maps of surfaces.

For example, consider the twistor fibration $\pi : J^+(S^4) \to S^4$. We may identify S^4 with $\mathbb{HP}^1$, the quaternionic projective line and $J^+(S^4)$ with $\mathbb{CP}^3$ so that the twistor fibration $\pi : \mathbb{CP}^3 \to \mathbb{HP}^1$ is the map sending a complex line in $\mathbb{C}^4 = \mathbb{H}^2$ to the quaternionic line generated by it. This is, of course, the celebrated Penrose fibration $\mathbb{CP}^3 \to S^4$. Now the J_1 complex structure on $\mathbb{CP}^3$ is just the usual one and the horizontal distribution is just the orthocomplement of the vertical distribution with respect to the Fubini-Study metric; for all of this see [Br2] and [E-Sa].

Now in the affine chart (z_1, z_2, z_3) the (1,0) part of the horizontal distribution is the kernel of the holomorphic 1-form given by

$$dz_1 - z_3 dz_2 + z_2 dz_3 \; .$$

Bryant shows that we may construct essentially all holomorphic integral curves on the horizontal distribution by the following recipe:

Theorem 5.1 [Br2] Let M be a connected Riemann surface and let f and g be meromorphic function on M with g non-constant. Define $\Phi(f,g) : M \to \mathbb{C}P^3$ by

$$\Phi(f,g) = [1, \; f - \tfrac{1}{2}g(df/dg), \; g, \; \tfrac{1}{2}(df/dg)].$$

Then $\Phi(f,g)$ is horizontal holomorphic. Further, if $\Phi : M \to \mathbb{C}P^3$ is horizontal holomorphic then it is of the form $\Phi(f,g)$ for a unique choice of meromorphic functions f,g on M or Φ has image in some $\mathbb{C}P^1 \subset \mathbb{C}P^3$.

Thus Theorem 4.1 provides harmonic maps of any Riemann surface into S^4.

The twistor spaces over a symmetric space G/K with integrable J_1 and holomorphic horizontal distribution have been classified by Bryant [Br3] and Salamon [S]. They are all of the form $G/H \to G/K$ discussed in Chapter III with (1,2) symplectic J_2.

In general the homogeneous twistor spaces $G/H \to G/K$ will not have holomorphic horizontal distribution, however there is always a sub-distribution of the horizontal distribution - 'a superhorizontal distribution' - which is G-invariant and holomorphic. By considering superhorizontal holomorphic maps into $G_2/U(2)$ (a twistor space over the exceptional symmetric space $G_2/SO(4)$), Bryant [Br1] was able to construct holomorphic maps into S^6 with its (1,2) symplectic non-integrable almost complex structure.

Horizontal holomorphic maps into twistor spaces are also relevant to the construction of Chapter II. Indeed, the isotropy conditions referred to therein are precisely the condition that a certain twistor lift be horizontal. This observation is the starting point for the (as yet far from complete) interpretation of the results and methods of Chapter II in the twistor framework.

B. *New harmonic maps from old*

Twistor methods often allow us to construct new harmonic maps from a given one:

Examples: (i) If we have a lift into a (1,2)-symplectic twistor space then that lift is harmonic by Theorem 1.1. Thus harmonic maps into Grassmannians satisfying a strong conformality condition produce harmonic maps into certain flag manifold as twistor lifts, cf. [B3], [I].

(ii) Sometimes a given almost complex manifold admits twistor fibrations over several Riemannian manifolds and so a holomorphic map projects onto several harmonic maps. As an example of this, consider the following set of homogeneous fibrations of flag manifolds:

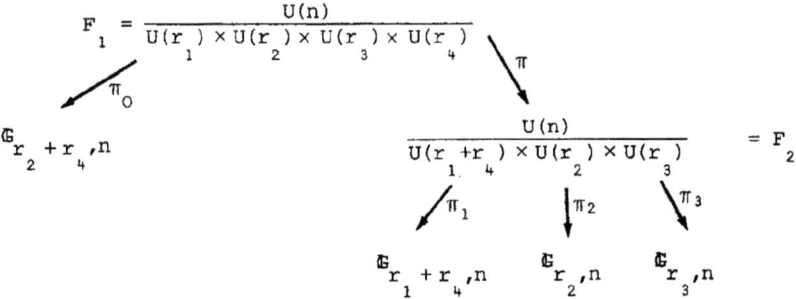

Here $n = r_1 + r_2 + r_3 + r_4$.

Now F_1 and F_2 admit almost complex structures for which π_i $i = 0,1,2,3$ and twistor fibrations and $\pi : F_1 \to F_2$ is holomorphic. Further F_2 is (1,2)-symplectic. Thus if ψ is a holomorphic map of a cosymplectic almost Hermitian manifold into F_1 we get four harmonic maps into Grassmannians and a harmonic map into F_2.

Now taking $r_1 = 1$, $n = 4$, results of Ramanathan [Ra] show that if $\phi : S^2 \to G_{2,4}$ is harmonic and non-±holomorphic then there is a holomorphic map $\psi : S^2 \to F_1$ with $\pi_o \circ \psi = \phi$. Thus we see that harmonic maps $S^2 \to G_{2,4}$ give rise to three other harmonic maps into Grassmannians.

For all these results and extensions thereof the reader is referred to [B1], [B2].

C. *An S^1-action on harmonic maps $S^2 \to G$, [U]*

Recall from Chapter IV, paragraph D, Uhlenbeck's twistor lift into the loop group. Given $\phi : S^2 \to G$, a harmonic map into a Lie group we built a map $\Phi : S^2 \to \Omega G$ by integrating a family of zero curvature equations.

Φ had the following properties:

(i) Φ was holomorphic and 'horizontal' in some sense

(ii) $\pi_{-1} \circ \Phi = \phi$ where π_{-1} is evaluation at -1.

Now S^1 acts on ΩG by rotation of the parameter

$$(e^{i\omega} \cdot \alpha)(e^{is}) = \alpha(e^{i(\omega+s)})(\alpha(e^{i\omega}))^{-1}$$

and this action preserves the complex structure and the horizontality condition on ΩG. Thus it acts by composition on the twistor lifts of harmonic maps $S^2 \to G$ and so by projection on the harmonic maps themselves. (Strictly we must normalise the value of ϕ to be the identity at some base point on S^2 to ensure uniqueness of the lift so that we do get a genuine S^1 action on harmonic maps).

Remarks (i) Using a similar, but more complicated, approach Uhlenbeck also constructs an action of the loop group itself on the space of harmonic maps $S^2 \to G$.

(ii) The fixed points of the S^1-action on ΩG are precisely the homomorphisms $S^1 \to G$. Thus the fixed points of the action on harmonic maps are those maps whose lifts take their image in a conjugacy class of homomorphisms G/H. Thus these maps have image in some copy of an inner Riemannian symmetric space embedded in G. In fact, the fixed points are precisely those harmonic maps into an inner Riemannian symmetric space G/K which are covered by a superhorizontal holomorphic map into a twistor space G/H.

REFERENCES

[A] AITHAL, A.R., Harmonic maps from S^2 to $\mathbb{C}_{2,5}$. Preprint, Tata Institute, Bombay, 1984.

[At] ATIYAH, M.F., Geometry of Yang-Mills fields, Fermi Lectures, *Sc. Norm. Sup. Pisa,* 1979.

[A-B] ATIYAH, M.F. and BOTT, R., The Yang-Mills equations over Riemann surfaces, *Phil. Trans. R. Soc. Lond. A,* **308** (1982), 523-615.

[B-P] BELAVIN, A.A. and POLYAKOV, A.M., Metastable states of two dimensional feromagnets, *JEP Lett.,* **22** (1975), 245.

[Br1] BRYANT, R.L., Submanifolds and special structures on the octonians, *J. Diff. Geom.,* **17** (1982), 185-232.

[Br2] BRYANT, R.L., Conformal and minimal immersions of compact surfaces into the 4-sphere, *J. Diff. Geom.,* **17** (1982), 455-473.

[Br3] BRYANT, R.L., Lie groups and twistor spaces. (To appear)

[B1] BURSTALL, F.E., Twistor fibrations of flag manifolds and harmonic maps of a 2-sphere into a Grassmannian. In: Proceedings on the Fifth International Colloquium in Differential Geometry, Santiago de Compostela, Pitman, London. (To appear)

[B2] BURSTALL, F.E., A twistor description of harmonic maps of a 2-sphere into a Grassmannian, Bath preprint 1985.

[B3] BURSTALL, F.E., Non-linear functional analysis and harmonic maps, Warwick thesis, 1984.

[B-W] BURSTALL, F.E. and WOOD, J.C., The construction of harmonic maps into complex Grassmannians, Leeds preprint 1985.

[Ca] CALABI, E., Minimal immersions of surfaces in Euclidean spheres, *J. Diff. Geom.,* **1** (1967), 111-125.

[C]   CHERN, S.S., Complex manifolds without Potential Theory, Springer-Verlag, Berlin, 1979.

[C-W]   CHERN, S.S. and WOLFSON, J.G., Harmonic maps of the 2-sphere into a complex Grassmannian manifold, II, Berkeley preprint 1985.

[D-Z]   DIN, A.M. and ZAKRZEWSKI, W.J., General classical solutions in the $\mathbb{CP}^{n-1}$ model, *Nucl. Phys.*, **B174** (1980), 397-406.

[EL1]   EELLS, J. and LEMAIRE, L., A report on harmonic maps, *Bull. Lond. Math. Soc.*, **10** (1978), 1-68.

[EL2]   EELLS, J. and LEMAIRE, L., Selected topics in harmonic maps, C.B.M.S. Regional Conference Series 50, American Mathematical Society, 1983.

[ESa]   EELLS, J. and SALAMON, S., Twistorial construction of harmonic maps of surfaces into four-manifolds, *Ann. Scuola Norm. Pisa.* (To appear)

[ES]   EELLS, J. and SAMPSON, J.H., Harmonic mappings of Riemannian manifolds, *Amer. J. Math.*, **86** (1964), 109-160.

[EW1]   EELLS, J. and WOOD, J.C., The existence and construction of certain harmonic maps, *Symp. Math. Roma*, **26** (1982), 123-138.

[EW2]   EELLS, J. and WOOD, J.C., Harmonic maps from surfaces to complex projective spaces, *Adv. in Math.*, **49** (1983), 217-263.

[ErW]   ERDEM, S. and WOOD, J.C., On the construction of harmonic maps into a Grasmannian, *J. Lond. Math. Soc.*, *(2)* **28** (1983), 161-174.

[GS]   GLASER, V. and STORA, R., Regular solutions of the $\mathbb{CP}^n$ models and further generalisations, CERN preprint, 1980.

[GH]   GRIFFITHS, P.A. and HARRIS, J., Principles of Algebraic Geometry, Wiley, 1978.

[Gr] GROMOV, M., Pseudo-holomorphic curves in symplectic manifolds, IHES preprint, 1985.

[G] GROTHENDIECK, A., Sur la classification des fibrés holomorphes sur la sphère de Riemann, *Amer. J. Math.*, **79** (1957), 121-138.

[G-O-R] GULLIVER, R.D., OSSERMAN, R. and ROYDEN, H., A theory of branched immersions of surfaces, *Amer. J. Math.*, **95** (1973), 750-812.

[H] HITCHIN, N.J., Monopoles and geodesics, *Comm. Math. Phys.*, **83** (1982), 579-602.

[I] ISHIHARA, T., The Gauss map and non-holomorphic harmonic maps, *Proc. Amer. Math. Soc.*, **89** (1983), 661-665.

[K-M] KOSZUL, J.L. and MALGRANGE, B., Sur certaines structures fibrées complexes, *Arch. Math.*, **9** (1958), 102-109.

[L] LAWSON, H.B., Lectures on minimal submanifolds I, Publish or Perish, 1980.

[Li] LICHNEROWICZ, A., Applications harmoniques et variétés Kählériennes, *Symp. Math. III,* Bologna (1970), 341-402.

[O-R] O'BRIAN, N. and RAWNSLEY, J.H., Twistor spaces, *Annals of Global Analysis and Geometry.* (To appear)

[P] PRESSLEY, A.N., The energy flow on the loop space of a compact Lie group, *J. Lond. Math. Soc.*, **26** (1982), 557-566.

[Ra] RAMANATHAN, J., Harmonic maps from S^2 to $\mathbb{C}_{2,4}$, *J. Diff. Geom.*, **19** (1984), 207-219.

[R] RAWNSLEY, J.H., f-structures, f-twistor spaces and harmonic maps, *Geometry Seminar Luigi Bianchi.* (To appear)

[S]   SALAMON, S., Harmonic and holomorphic maps, *Geometry Seminar Luigi Bianchi*. (To appear)

[U]   UHLENBECK, K.K. Harmonic maps into Lie groups (Classical solutions of the chiral model), Preprint, University of Chicago, 1985.

[W]   WELLS, R.O., Differential analysis on complex manifolds, *Grad. Text. Math.,* 65 Springer-Verlag, New York, 1980.

TWISTOR METHODS

by

John Rawnsley

Preface

In these notes I shall describe a setting for twistor methods in Riemannian Geometry. I shall not look at Penrose's original theory for Lorentzian 4-manifolds which is a huge subject in its own right (see Wells [1979] for references). I do not claim the most general or definitive formulation as this area is still undergoing developement. The definition of twistor space I have chosen is sufficiently general to have interesting non-trivial examples and sufficiently special to be capable of analysis in a few lectures. Other kinds of twistor spaces have already been used and will doubtless be of importance in the future. To illustrate the theory I have chosen the construction of self-dual Yang-Mills fields on S^4 by Atiyah, Drinfeld, Hitchin, Manin [1978] because of its relevance to the lectures of Professor Bourguignon and Professor Taubes. Applications to harmonic maps will be the subject of the series of lectures by Professor Burstall.

Lecture I: Introduction

Although the term twistor methods has only appeared recently in differential geometry, techniques in the same spirit have long been used as we shall see. The name twistor comes from the work of Roger Penrose in relativity. Robinson [1961] found a correspondence between null solutions for the Minkowski metric of Maxwell's equations and certain null congruences of geodesics. These congruences are shear-free but have twist and correspond with 3-dimensional submanifolds of a certain hypersurface in CP^3. The C^4 of which CP^3 is the projective space is the space of spinors of Minkowski space. Combining twist and spinor into a single word produces the name twistor for this C^4 and projective twistor space for CP^3. Penrose [1967] discovered he could in fact obtain all real-analytic solutions of Maxwell's equations by means of contour integrals of suitable functions defined on open sets in CP^3 and by using more general integrands he obtained also solutions to zero rest-mass field equations for arbitrary spins. Atiyah later pointed out that the large degree of freedom in the choice of the contour and integrand meant that the invariantly defined object on CP^3 was really an element of a suitable (Dolbeault or sheaf) cohomology group.

The above situation thus takes a problem from the real differential geometric domain and translates it into a problem in complex analysis. The latter is much more highly developed and has a very rich structure which can be used to solve the original problem. By a twistor method we shall mean such a translation and will not limit the use of the term to its original highly specific 4-dimensional context. In particular the building blocks for our twistor method need not be related to spinors at all.

The geometrical relationship of CP^3 with the original Minkowski space is by means of a double fibration:

$$\begin{array}{ccc} & F & \\ \pi \swarrow & & \searrow \sigma \\ CP^3 & & M^C \end{array}$$

Double fibrations generalizing this one to higher dimensions have been studied by Penkov [1984]. He shows how various sheaves on the twistor space give rise to differential equations on the manifold.

Twistor methods have also been used in Riemannian geometry for more than a century. In many cases the situation is simpler than in the indefinite Lorentzian signature in that there is a complex manifold Z (the twistor space for the problem) fibring directly over the manifold M being studied:

$$\pi : Z \longrightarrow M$$

and the fibres are usually (but not always) complex submanifolds. It is in this context which I shall mostly be discussing twistor methods in these lectures. However, to provide a simple example of a twistor method in Riemannian geometry I shall first describe a problem studied by Weierstrass [1866]. The description given here is from Lawson's book [1980] but see also Hitchin [1982].

If $D \subset C$ is a domain, a map $\phi : D \longrightarrow R^n$ is harmonic if each component ϕ_i is a harmonic function and conformal if $d\phi$ preserves angles in the tangent spaces. It can be seen that where ϕ is an immersion, then being conformal and harmonic makes $\phi(D)$ a minimal surface and conversely. Since only branch-points can occur and they do not cause any difficulty, the problem of studying minimal surfaces in R^n becomes one of studying conformal harmonic maps. The condition for harmonicity can be written in terms of a complex coordinate z on D as

$$\frac{\partial^2 \phi}{\partial \bar{z} \partial z} = 0.$$

If $\psi = \partial \phi / \partial z : D \longrightarrow C^n$, then the above equation says ψ is a vector of holomorphic functions. ϕ is conformal if and only if

(1) $$\sum_i \psi_i^2 = 0.$$

If we make a holomorphic change of coordinate $w = w(z)$, then

$$\frac{\partial \phi}{\partial z} = \frac{dw}{dz} \frac{\partial \phi}{\partial w} \quad \text{or} \quad \psi = \frac{dw}{dz} \tilde{\psi}.$$

Thus the line $[\psi]$ spanned by ψ is independent of the coordinates used and

$$\tilde{\phi} = [\psi]$$

gives a map into CP^{n-1}. Taking into account the conformality condition (1), we see $\tilde{\phi}$ is actually a map into the quadric Q_{n-2}:

$$\tilde{\phi} : D \longrightarrow Q_{n-2}.$$

$\tilde{\phi}$ is of course holomorphic. Since it is defined in a coordinate independent way, it can be pieced together from charts on a general Riemann surface. The complex conjugate of $\tilde{\phi}$ is the Gauss map of ϕ after identifying Q_{n-2} with $G_2^+(R^n)$. Thus the Gauss map of a minimal surface is antiholomorphic.

For $n = 3$ this translation of the problem into a holomorphic one has a completely explicit solution as we now see. If ϕ is non-constant, $\psi \neq 0$, so we can choose a basis for C^3 so that $\psi_3 \neq 0$. Then $\psi_1 \neq i\psi_2$ and

(2) $$g = \psi_3/(\psi_1 - i\psi_2)$$

is a meromorphic function on M independent of coordinates. Likewise

(3) $$w = \psi_3 dz$$

is an invariantly defined holomorphic differential on M. Conversely, given a holomorphic differential $w = fdz$ and a meromorphic function g then (1-3) can be solved to give

$$\psi_1 = \tfrac{1}{2}(1-g^2)f, \quad \psi_2 = \tfrac{1}{2}i(1+g^2)f, \quad \psi_3 = f.$$

These will have no poles if f has a zero of order 2m wherever g has a pole of order m. This is the only constraint locally.

Example

w = dz, g = z gives

$$\psi_1 = \tfrac{1}{2}(1-z^2), \quad \psi_2 = \tfrac{1}{2}i(1+z^2), \quad \psi_3 = 1$$

and integrating

$$\phi_1 = \text{Re}(z-z^3/3), \quad \phi_2 = \text{Rei}(z+z^3/3), \quad \phi_3 = \text{Rez}.$$

This map is Enneper's surface in R^3.

The following are some more examples of problems which have been solved by twistor methods. This list is not complete, but gives some idea of the scope of these methods. I apologize to anyone I have inadvertantly not mentioned.

1/ Calabi [1967] studied minimal 2-spheres in S^{2n} by a method analogous to the Weierstrass representation and converted them into holomorphic curves in a flag manifold.

2/ Atiyah and Ward [1977] translated the problem of finding self-dual solutions of the Yang-Mills equations on S^4 into finding certain holomorphic vector bundles on CP^3, a problem which was then solved using results from algebraic geometry by Atiyah, Drinfeld, Hitchin and Manin [1978]. This will constitute the topic of the last lecture in my course.

3/ Atiyah, Hitchin and Singer [1978] generalized example 2 to a class of 4-manifolds with self-dual Riemannian curvature. Cohomology groups on the twistor space correspond with solutions of differential equations generalizing the zero rest-mass equations. Penkov [1984] has given a general setting for such correspondences.

4/ Berard-Bergery [1979] and Salamon [1980] found twistor spaces for manifolds admitting quaternionic structures and used these complex manifolds to study the quaternionic structures.

5/ Hitchin [1979] constructed certain Einstein metrics on 4-manifolds by first constructing a twistor space. This method has also been used by Poon [1985] and Pedersen [1985].

6/ Hitchin [1982] translated the Bogomolny equations for non-abelian magnetic monopoles into the problem of finding certain holomorphic curves in an associated complex surface. See also Murray [1983].

7/ Eells and Wood [1983] classified all minimal 2-spheres in CP^n in terms of holomorphic curves by a method analogous to Calabi's in 1/. Their methods work also for surfaces of non-zero genus in some cases and complete partial results by Burns [1982], Din and Zakrzewski [1980] and Glaser and Stora [1980] obtained in studying certain non-linear fields (σ-models). This will be described in detail in Burstall's lectures as well as general results of this kind for harmonic maps due to Burstall, Eells, Rawnsley, Salamon and Wood for more general target manifolds. This is work still in progress, and uses a more general kind of twistor space since the really significant structure in this case turns out to be a non-integrable almost-complex structure found by Eells and Salamon [1983]. Twistor methods have also been used by Bryant [1982] to prove the existence of minimal surfaces in S^4.

8/ Recently Uhlenbeck [1985] has shown how to lift minimal 2-spheres in a compact Lie group G into a holomorphic curve in the based loop space ΩG. Burstall has shown that the loop space (a Kaehler manifold) fibres over the group and has an almost complex structure with the property that any holomorphic curve in ΩG projects to a minimal surface in G. This provides an example of a twistor space whose fibres are not complex submanifolds.

9/ The full Yang-Mills equations on S^4 have been given a twistor translation by Witten [1978] and Green, Isenberg, Yasskin [1978]. The twistor space has a structure sheaf with nilpotent elements and the translation has not yet proven of use in solving the equations.

Lecture II: The Universal Twistor Space

The kind of twistor space I am going to study in these lectures is typified by a complex manifold Z fibred submersively over a manifold M

$$\pi : Z \longrightarrow M$$

with the fibres complex submanifolds. We recall from Professor Green's lectures that the complex structure of Z can be described by an endomorphism J of the tangent bundle TZ which satisfies $J^2 = -1$. At $p \in Z$ we denote by V_p the tangent space to the fibre through p (or the vertical space at p) which consists of

$$\{ X \in T_p Z : \pi_* X = 0 \}.$$

The fibres will be complex submanifolds provided V_p is J-stable for each p:

$$JV_p \subset V_p,$$

and then the restriction of J to the fibre gives the complex structure of each fibre.

If $x = \pi(p)$, then π_* maps $T_p Z$ onto $T_x M$ with kernel V_p, so

$$T_x M = T_p Z / V_p .$$

It follows that there is an endomorphism $j(p)$ of $T_x M$ with $j(p)^2 = -1$ induced by J. This endomorphism will in general vary as p moves in the fibre $\pi^{-1}(x)$.

This observation suggests how we might try to build twistor spaces by taking families $\{j(p)\}_{p \in Z_x}$ of endomorphisms of $T_x M$ of square -1. If Z_x is complex and we can lift each $j(p)$ horizontally at $p \in Z = \bigcup_x Z_x$, then we can fit the vertical and horizontal endomorphisms together to give an almost complex structure on Z. It may not be integrable, but if it is we shall have a possible twistor space related to the manifold M. In this lecture we shall look at this integrability question in a context sufficiently general to provide a lot of the known examples of twistor spaces. We begin with what can be regarded as the universal example.

If we identify C^n with R^{2n} by

$$z = x + iy \mapsto \begin{bmatrix} x \\ y \end{bmatrix},$$

then

$$iz = -y + ix \mapsto \begin{bmatrix} -y \\ x \end{bmatrix} = \begin{bmatrix} 0 & -1 \\ 1 & 0 \end{bmatrix}\begin{bmatrix} x \\ y \end{bmatrix}$$

so the complex structure of R^{2n} as C^n is determined by

$$J_o = \begin{bmatrix} 0 & -1 \\ 1 & 0 \end{bmatrix}.$$

But R^{2n} has many more complex vector space structures. We denote by $\underline{J}_n$ the set of all endomorphisms J of J^{2n} with $J^2 = -1$. Each of these is a possible complex structure such that J gives the action of $i \in C$. If $g \in GL(2n,R)$ then $gJ_o g^{-1}$ is in $\underline{J}_n$. In fact this gives all of $\underline{J}_n$ and so the latter is a homogeneous space. The elements of $GL(2n,R)$ fixing J_o are the complex linear maps of C^n, so form a subgroup of $GL(2n,R)$ equal to $GL(n,C)$. Hence

$$\underline{J}_n = GL(2n,R)/GL(n,C)$$

as a homogeneous space.

This gives $\underline{J}_n$ a smooth manifold structure whose tangent space at J_o will be the quotient

$$g\ell(2n,R)/g\ell(n,C)$$

of the Lie algebras. $gl(2n,R)$ consists of all $2n \times 2n$ matrices and $gl(n,C)$ those which commute with J_o. In fact if $A+iB \in gl(n,C)$, then

$$(A+iB)(x+iy) = Ax-By+i(Bx+Ay) \mapsto \begin{bmatrix} Ax-By \\ Bx+Ay \end{bmatrix} = \begin{bmatrix} A & -B \\ B & A \end{bmatrix} \begin{bmatrix} x \\ y \end{bmatrix}$$

so $gl(n,C)$ sits inside $gl(2n,R)$ as all matrices with the block form

$$\begin{bmatrix} A & -B \\ B & A \end{bmatrix}.$$

Any matrix $\xi \in gl(2n,R)$ can be decomposed

$$\xi = \tfrac{1}{2}(\xi - J_o \xi J_o) + \tfrac{1}{2}(\xi + J_o \xi J_o).$$

It is easy to check that the first part commutes with J_o and the second part anticommutes with J_o. If we denote by $\underline{m}$ the set of all elements of $gl(2n,R)$ which anticommute with J_o then we see we have a direct sum decomposition

$$gl(2n,R) = gl(n,C) + \underline{m}.$$

It follows we can identify $\underline{m}$ with the tangent space $T_{J_o} J_n$.

Consider the following operation

$$\xi \mapsto J_o \xi$$

on $\underline{m}$. Obviously

$$J_o(J_o \xi) = -\xi$$

so left multiplication by J_o on $\underline{m}$ gives $T_{J_o} J_n$ a complex structure. This works in fact for every tangent space:

$$T_J J_n = \{ \xi \in gl(2n,R) : \xi J = -J\xi \}$$

and left multiplication by J gives a complex structure to the tangent space at J. Thus $\underline{J}_n$ carries a canonical almost complex structure. It is in fact integrable.

One way to see this last fact is to describe $\underline{J}_n$ in a manifestly complex way. Any J in $\underline{J}_n$ extends linearly to $\mathbb{C}^{2n}$ satisfying still $J^2 = -1$ and so has two eigenvalues $\pm i$. Since J is real the eigenspaces are conjugates, say W and $\overline{W}$ and $\mathbb{C}^{2n}$ is the direct sum $W + \overline{W}$. In fact given a complex subspace W of $\mathbb{C}^{2n}$ of dimension n with $W \cap \overline{W} = 0$ then W is the $+i$ eigenspace of a unique J in $\underline{J}_n$. Thus we can identify $\underline{J}_n$ with an open set in the complex Grassmannian $G_n(\mathbb{C}^{2n})$ and so give $\underline{J}_n$ this complex structure. It is an exercise to show that the two complex structures I have given to $\underline{J}_n$ are in fact the same.

If we have a manifold M of dimension $2n$, let $\pi_o : E \longrightarrow M$ be the $GL(2n,R)$ bundle of all bases of the tangent bundle TM. Any such basis b can be viewed as a linear isomorphism

$$b : R^{2n} \longrightarrow T_xM, \qquad x = \pi_o(b).$$

Then we can form the associated bundle

$$E \times_{GL(2n,R)} \underline{J}_n$$

with fibre $\underline{J}_n$. It can be viewed in several ways. If $J \in \underline{J}_n$ and $b \in E$ then bJb^{-1} is an endomorphism of T_xM with square -1. If we let $J(M)$ denote the bundle whose fibre at x is all such endomorphisms of T_xM, then we see

$$J(M) = E \times_{GL(2n,R)} \underline{J}_n.$$

There is a map

$$\pi_1 : E \longrightarrow J(M)$$

which is surjective given by

$$\pi_1(b) = bJ_o b^{-1}$$

and clearly has fibres the orbits of the subgroup $GL(n,C)$:

$$J(M) = E/GL(n,C)$$

and E is a principal $GL(n,C)$ bundle over $J(M)$. $J(M)$ is our universal twistor space because we saw at the beginning of this lecture that any twistor space Z has a map

$$j : Z \longrightarrow J(M).$$

We now try to make $J(M)$ into a complex manifold in such a way that j has the best chance to be holomorphic. To find such a complex structure we begin by observing that the complex structure of $J_{\underline{n}}$ transfers to every fibre of $J(M)$, so that $J(M)$ is a bundle of complex manifolds over M and we have the corresponding endomorphism J^v of the vertical tangent bundle V. A connection ∇ in TM gives a horizontal distribution on E which projects by π_1 to a horizontal distribution H^∇ on $J(M)$. Each space H^∇_J is mapped isomorphically by π_* to T_xM if $x = \pi(J)$ and so J can be lifted horizontally at J to act on H^∇_J as $\hat{J}$ say. Defining J^h at J as $\hat{J}$ gives H^∇ a complex structure. Since $TJ(M)$ is the direct sum of V and H^∇, J^v and J^h fit together to give an almost complex structure J_1 on $J(M)$. The only ingredient used to build it was the connection ∇. J_1 depends on the choice of the connection, but for convenience we shall not indicate this in the notation so long as no confusion arises. The suffix 1 appears on J_1 because there is more than one way to fit J^v and J^h together. In harmonic map theory a second choice $J_2 = -J^v+J^h$ is even more important than J_1, but J_2 is not integrable.

To study the integrability question for J_1 we use the Nirenberg-Newlander theorem which tells us that an almost complex structure is integrable provided the 1,0 vectors are closed under Lie brackets, or dually, if the ideal of all differential forms generated by the 1,0 forms is closed under the exterior derivative d. This last formulation can be checked by pulling the forms back to E by π_1. On E we have certain natural 1-forms spanning the cotangent spaces. First we have the connection form ω of the connection ∇ and second we have the soldering form θ which is the R^{2n}-valued 1-form defined by

$$\theta_b = b^{-1} \circ \pi_{o*}.$$

The components of ω and θ are independent and $4n^2+2n$ in number which is the dimension of E as a manifold. We need a Lemma:

Lemma If $P^{\pm}$ denote the projections onto the $\pm i$ eigenspaces of J_- on C^{2n} then

$$P^+\theta, \quad P^+\omega P^-$$

span the pull-back to E of the 1,0 forms for J_1.

Proof $X \in T_b E$ projects to a 0,1 vector on $J(M)$ under π_{1*} provided $\pi_{o*}X$ is 0,1 for $bJ_o b^{-1}$ and the part of $\omega(X)$ in $\underline{m}$ satisfies

$$J_o \omega(X)\underline{m} = -i\omega(X)\underline{m}.$$

This last condition is easily seen to be

$$P^+\omega(X)P^- = 0$$

whilst the first reads

$$bJ_o b^{-1}\pi_{o*}X = -i\pi_{o*}X$$

or

$$P^+\theta(X) = 0.$$

Thus $P^+\theta$ and $P^+\omega P^-$ are 1,0 forms and a dimension count shows their components span.

Corollary J_1 is integrable if and only if $P^+\theta$ and $P^+\omega P^-$ generate a d-closed ideal.

To translate this statement into geometrical terms we use the torsion and curvature of ∇. The torsion T is a 2-form with values in TM and lifts to E as the R^{2n}-valued 2-form τ given by

$$\tau_b(X,Y) = b^{-1}T_x(\pi_{0*}X, \pi_{0*}Y), \qquad x = \pi_0(b).$$

Then

$$\tau(X,Y) = (d\theta)(X^h, Y^h) = (d\theta + \omega \wedge \theta)(X,Y).$$

Similarly the curvature R of ∇ in TM lifts to the $g\ell(2n,R)$-valued 2-form Ω on E:

$$\Omega_b(X,Y) = b^{-1}R_x(\pi_{0*}X, \pi_{0*}Y)b$$

and

$$\Omega(X,Y) = (d\omega)(X^h, Y^h) = (d\omega + \tfrac{1}{2}[\omega \wedge \omega])(X,Y).$$

Thus we can write

$$d\theta = \tau - \omega \wedge \theta, \quad d\omega = \Omega - \tfrac{1}{2}[\omega \wedge \omega].$$

Then

$$d(P^+\theta) = P^+\tau - P^+\omega \wedge \theta = P^+\tau - P^+\omega P^- \wedge \theta - P^+\omega \wedge P^+\theta,$$

and

$$d(P^+\omega P^-) = P^+\Omega P^- - \tfrac{1}{2}P^+[\omega P^- \wedge \omega]P^- - \tfrac{1}{2}P^+[\omega \wedge P^+\omega]P^-.$$

Hence $dP^+\theta$ and $dP^+\omega P^-$ are congruent modulo the ideal generated by 1,0 forms to $P^+\tau$ and $P^+\Omega P^-$. Hence we now have

Theorem 1 J_1 on $J(M)$ is integrable if and only if $P^+\tau$, $P^+\Omega P^-$ are in the ideal generated by 1,0 forms.

This formulation can be simplified somewhat by observing that a form is in the ideal if it vanishes when all its arguments are 0,1 vectors. If $J \in J(M)$ let $J^\pm$ be the corresponding projections onto the $\pm i$ eigenspaces, so J^-X are the 0,1 vectors of J on T_xM^C. Then if X, Y are 0,1 vectors we have $J = bJ_ob^{-1}$ and

$$0 = P^+\tau_b(X,Y) = P^+b^{-1}T_x(\pi_{0*}X,\pi_{0*}Y) = b^{-1}J^+T_x(\pi_{0*}X,\pi_{0*}Y)$$

hence

$$J^+T_x(J^-X,J^-Y) = 0, \quad \forall\ J \in J(M)_x.$$

A similar calculation for the curvature shows

$$J^+R_x(J^-X,J^-Y)J^- = 0, \quad \forall\ J \in J(M)_x.$$

Thus we have the revised form of the integrability condition:

Theorem 1' J_1 on $J(M)$ is integrable if and only if for every J in $J(M)$,

$$J^+T(J^-X,J^-Y) = 0, \quad J^+R(J^-X,J^-Y)J^- = 0.$$

Lecture III: The Integrability Condition

In this lecture we study the integrability conditon on the torsion and cuvature (Theorem 1, 1') we obtained last time. The key to their solution is to translate them into a problem in representation theory.

If s is a section of an associated bundle

$$E \times_{GL(2n,R)} F$$

with fibre F, it can be identified with a map $\hat{s}: E \longrightarrow F$ by

$$\hat{s}(b) = b^{-1} s(\pi_0(b)),$$

viewing $b \in E$ as an isomorphism of the standard fibre F with the fibre at $\pi_0(b)$. $\hat{s}$ will satisfy

$$\hat{s}(b.g) = g^{-1}.\hat{s}(b), \quad \forall\ g \in GL(2n,R).$$

The tension field T is a section of the bundle with fibre $W \otimes \Lambda^2 W^*$, where $W = R^{2n}$, so it corresponds with a map

$$\hat{T}: E \longrightarrow W \otimes \Lambda^2 W^*$$

transforming under the natural action of $GL(2n,R)$ on $W \otimes \Lambda^2 W^*$. We have

$$\hat{T}(b)(u,v) = b^{-1} T_{\pi_0(b)}(bu,bv),$$

so the integrability condition at $J = bJ_0 b^{-1}$ becomes

(4) $\qquad P^+ \hat{T}(b)(P^- u, P^- v) = 0.$

It follows that $\hat{T}$ has values in the largest $GL(2n,R)$-invariant subspace of $W \otimes \Lambda^2 W^*$ satisfying (4). (4) is not an invariant condition so it needs some care in use. It admits a representation theoretic translation as follows:

If $T \in W \otimes \Lambda^2 W^*$ it is acted on by $g \in GL(2n,R)$ by

$$(g.T)(u,v) = gT(g^{-1}u, g^{-1}v).$$

This action can be differentiated to yield a representation of the Lie algebra $g\ell(2n,R)$ on $W \otimes \Lambda^2 W^*$

$$(\xi.T)(u,v) = \xi T(u,v) - T(\xi u, v) - T(u, \xi v).$$

If we view J_o as an element of $g\ell(2n,R)$, it acts on $W \otimes \Lambda^2 W^*$ by this differentiated action and has eigenvalues $\pm 3i$, $\pm i$. The projection $\mathcal{P}$ onto the $3i$ eigenspace is easily seen to be

$$(\mathcal{P}T)(u,v) = P^+ T(P^- u, P^- v).$$

Hence (4) can be reinterpreted as saying that $\hat{T}$ has values in the largest $GL(2n,R)$-invariant subspace of $W \otimes \Lambda^2 W^*$ disjoint from the $3i$ eigenspace of J_o.

A similar interpretation is possible for the curvature condition. We define

$$\hat{R} : E \longrightarrow \text{End} W \otimes \Lambda^2 W^*$$

by

$$\hat{R}(b)(u,v) = b^{-1} R_{\pi_0(b)}(bu, bv) b,$$

which has the appropriate transformation property under the action

$$(g.R)(u,v) = gR(g^{-1}u, g^{-1}v)g^{-1}$$

on elements of $\text{End} W \otimes \Lambda^2 W^*$. The integrability condition at $J = bJ_o b^{-1}$ becomes

(5) $\quad P^+\hat{R}(b)(P^-u,P^-v)P^- = 0.$

Differentiating the action of $GL(2n,R)$ gives

$$(\xi.R)(u,v) = \xi R(u,v) - R(\xi u,v) - R(u,\xi v) - R(u,v)\xi$$

and for $\xi = J_o$ condition (5) says that J_o has no $4i$ eigenspace in common with the values of $\hat{R}$. The possible eigenvalues in this representation are $\pm 4i$, $\pm 2i$, 0, so the integrability condition rules out the largest potential eigenvalue for the curvature just as it did for the torsion. We thus have a third formulation of the theorem:

<u>Theorem 1"</u> J_1 on $J(M)$ is integrable if and only if the torsion $\hat{T}$ and curvature $\hat{R}$ have values in the largest $GL(2n,R)$-invariant subspaces of $W \otimes \Lambda^2 W^*$ and $EndW \otimes \Lambda^2 W^*$ (respectively) on which J_o has no $3i$ or $4i$ eigenvalues in the Lie algebra representation.

This is the form of the integrability condition which is most easy to apply in practice. The idea is to analyse the spaces $W \otimes \Lambda^2 W^*$ and $EndW \otimes \Lambda^2 W^*$ into irreducible pieces and check the possible eigenvalues of J_o. Any time it has a $3i$ or $4i$ eigenvalue, that whole component of $\hat{T}$ of $\hat{R}$ must vanish. Let us illustrate this process on $\hat{T}$.

$W \otimes \Lambda^2 W^*$ splits into two irreducible pieces, the kernel of the contraction

$$W \otimes \Lambda^2 W^* \longrightarrow W^*$$

and a complementary subspace which is a copy of W^*. On W^* J_o has eigenvalues $\pm i$, so on the other subspace the $3i$ must occur and hence that component is zero. Thus $\hat{T}$ must have values in $W^* \subset W \otimes \Lambda^2 W^*$. Converted into a statement about T on M, this means there is a section β of $E^\times{}_{GL(2n,R)} W^* = T^*M$ - that is β is a 1-form - such that

$$T(X,Y) = \beta(X)Y - \beta(Y)X.$$

Consider a new connection

$$\nabla'_X Y = \nabla_X Y - \beta(X)Y.$$

It has torsion T' given by

$$T'(X,Y) = T(X,Y) - \beta(X)Y - \beta(Y)X$$

$$= 0.$$

∇' will define an almost complex structure J'_1 on $J(M)$ whose 1,0 forms on E are spanned by $P^+\theta$, $P^+\omega'P^-$ where ω' is the connection form for ∇'. But

$$\omega' = \omega - \pi_0^*\beta.1$$

so

$$P^+\omega'P^- = P^+\omega P^- - \pi_0^*\beta P^+P^- = P^+\omega P^-$$

since $P^+P^- = 0$. It follows $J'_1 = J_1$ so integrability implies there is a torsion-free connection defining J_1. Without loss of generality we can now assume we are dealing with this connection.

The curvature condition can be discussed in the same manner. Since ∇ is now torsion-free its curvature R satisfies the first Bianchi identity

$$R(X,Y)Z + R(Y,Z)X + R(Z,X)Y = 0.$$

Tensors of curvature type satisfying this condition form a subspace of $EndW \otimes \Lambda \hat{W}^*$ and this is the subspace we must analyse into irreducibles. We can contract out a copy of $\Lambda^2 W^*$ and of $S^2 W^*$ and what remains is irreducible and certainly contains a 4i eigenvalue since none of the small representations we get by contraction does. So R has values in $\Lambda^2 W^* \oplus S^2 W^* = W^* \otimes W^*$.

This means there is a bilinear form $\mu \in C^\infty(T^*M \otimes T^*M)$ such that if we form

$$\tilde{\mu}(X,Y)Z = \mu(X,Y)Z - \mu(Y,X)Z + \mu(X,Z)Y - \mu(Y,Z)X$$

then

(6) $\qquad R = \tilde{\mu}.$

The second Bianchi identity imposes also a derivative condition on μ, namely

$$(\nabla_X \mu)(Y,Z) = (\nabla_Y \mu)(X,Z).$$

Thus we have the integrability condition for J_1: TM admits a torsion-free connection defining J_1 whose curvature has the special form (6). This is a highly restrictive condition (projectively flat) and instead of analyzing it further we shall look at how to construct other examples in which there may be a better chance for J_1 to be integrable. The basic problem is that the fibre of $J(M)$ is too big so (4) and (5) impose too many conditions. We try to find similar bundles with smaller fibres.

The results of this lecture and the next are from Dubois-Violette [1981], Berard-Bergery and Ochiai [1982], O'Brian and Rawnsley [1985] and the 4-dimensional example is of course the motivating one of Atiyah, Hitchin and Singer [1978].

Lecture IV: Twistor Spaces

Suppose we have a bundle

$$\pi : Z \longrightarrow M$$

with complex fibres, a fibre-preserving map

$$j : Z \longrightarrow J(M)$$

and a horizontal distribution H on Z so that $TZ = H \oplus V$. V carries an endomorphism J^V giving the complex structures of the fibres and we can lift j(z) horizontally at $z \in Z$ using the isomorphism

$$\pi_* : H_z \longrightarrow T_{\pi(z)} M.$$

This defines the endomorphism J^h on H and so J_1 on TZ:

$$J_1 = J^h \oplus J^V.$$

We can ask if J_1 is integrable on Z just as for J(M). In this generality it is difficult to say anything. If however we make the following two assumptions

(7i)   j is holomorphic on each fibre,

(7ii)   there is a connection ∇ in TM and $j_* H \subset H^\nabla$,

then j is holomorphic in the sense that its differential j_* intertwines J_1 on Z with that on J(M). It now follows that the obstruction to integrability (the Nijenhuis tensor) on J(M) must vanish on $j_* TZ$ when J_1 is integrable on Z. Thus the torsion and curvature of ∇ satisfy

(8) $J^+ T(J^- X, J^- Y) = 0, \quad J \in j(Z),$

(9) $J^+ R(J^- X, J^- Y) J^- = 0, \quad J \in j(Z).$

Thus the integrability condition now only need hold on $j(Z)$ and not necessarily for the whole of $J(M)$. Thus we have

Theorem 2 Suppose Z is a twistor space over M satisfying (7i, ii) for some map j and connection ∇ then J_1 on Z is integrable only if (8) and (9) hold for the complex structures in $j(Z)$.

In general these conditions are only necessary, but if j is an immersion they are also sufficient.

We can get examples satisying (7i,ii) as follows: Suppose M has a G-structure. By this we mean G is a Lie group with a representation on $W = R^{2n}$ and there is a principal G-bundle P over M with an isomorphism

$$TM \cong P \times_G W.$$

Since G acts on R^{2n} it also acts on $\underline{J}_n$ and suppose $F \subset \underline{J}_n$ is a G-invariant complex submanifold. The inclusion map $F \hookrightarrow \underline{J}_n$ induces a map

$$j : P \times_G F \longrightarrow P \times_G \underline{J}_n = J(M)$$

which is holomorphic on each fibre. Choosing a connection ω on P gives a covariant derivative ∇ in TM and horizontal distributions on $Z = P \times_G F$ and $J(M)$ which necessarily satisfy (7ii).

The torsion and curvature can be viewed as functions on P with values in a representation of G and which are equivariant for G. The previous argument can then be applied to translate the integrability conditions into a problem in representation theory for G. We decompose $W \otimes \Lambda^2 W^*$ and $EndW \otimes \Lambda^2 W^*$ into G-irreducibles and then inspect each component for $3i$ or $4i$ eigenvalues of J_o.

The simplest example of a G-structure is given by a Riemannian metric on M. Then $G = O(2n)$ and P is the bundle of orthonormal frames. If we take F to be all J in $\underline{J}_n$ compatible with the metric we get a bundle we denote by $J(M,g)$. The Levi-Civita connection gives a torsion-free connection on P and if we use this to put J_1 on $J(M,g)$ then the integrability condition reduces to a condition on the Riemannian curvature tensor R. Namely that R must take its values in the subspace of the space of curvature tensors orthogonal to the ireducible subspaces where J_o has a 4i eigenvalue.

Under the action of $O(2n)$ the space of curvature tensors with the same symmetries as the Riemannian curvature breaks into three irreducible pieces corresponding with the scalar curvature, the traceless Ricci tensors and the Weyl curvature tensors. The scalar curvature and Ricci curvature are in representations which are too small to contain any 4i eigenvalue for J_o, so this must occur on the Weyl tensors. Hence integrability in this case is equivalent to the vanishing of the Weyl tensor.

<u>Theorem 3</u> J_1 on $J(M,g)$ defined by the Levi-Civita connection is integrable if and only if g is locally conformal to a flat metric.

<u>Remarks</u> 1. It is possible to begin with any metric connection in P. The torsion condition then forces it to define the same almost complex structure as the Levi-Civita connection. The argument is essentially the same as for $J(M)$.

2. $J(M,g)$ and its almost complex structure can easily be seen to depend only on the conformal class of g, so it is not surprising that the integrability condition depends only on the conformally invariant part of the curvature - the Weyl tensor.

The next structure we consider is to add an orientation to the metric. Thus we consider an oriented Riemannian manifold (M,g) and its oriented orthonormal frame bundle P. The latter is a principal $SO(2n)$ bundle. For F we take the metric compatible J's in $\underline{J}_n$ whose complex orientation coincides with the given orientation. The associated bundle we denote by $J_+(M,g)$. It consists of all orientation and metric compatible complex structures on the tangent spaces of M. If we use the Levi-Civita connection to define J_1 then for dim $M \geq 6$ the answer is the same as before.

<u>Theorem 4</u> J_1 on $J_+(M,g)$ is integrable for dim $M \geq 6$ if and only if g is locally conformally flat.

In case M has dimension 4, 4i still occurs in the Weyl tensors, but they are no longer irreducible under $SO(4)$. The Weyl tensor splits into two pieces

$$W = W_+ + W_-$$

where

$$*W_+ = W_+, \quad *W_- = -W_-$$

and $*$ is the Hodge duality operator. In fact 4i only occurs in one of these pieces. To see which it is we look at the decomposition

$$\Lambda^2 W = \Lambda^2_+ + \Lambda^2_-$$

into self-dual and anti-self-dual parts. If we choose an oriented basis e_1, e_2, e_3, e_4 for W so that

$$J_0 e_1 = e_2, \quad J_0 e_3 = e_4,$$

then Λ^2_+ is spanned by

$$\alpha_1^+ = e_2 \wedge e_3 + e_1 \wedge e_4, \quad \alpha_2^+ = e_3 \wedge e_1 + e_2 \wedge e_4, \quad \alpha_3^+ = e_1 \wedge e_2 + e_3 \wedge e_4.$$

and Λ^2_- by

$$\alpha_1^- = e_2 \wedge e_3 - e_1 \wedge e_4, \quad \alpha_2^- = e_3 \wedge e_1 - e_2 \wedge e_4, \quad \alpha_3^- = e_1 \wedge e_2 - e_3 \wedge e_4.$$

It is easy to work out how J_0 acts in the Lie algebra representation on these. For example

$$J_0 \alpha_1^+ = -e_1 \wedge e_3 + e_2 \wedge e_4 + e_2 \wedge e_4 - e_1 \wedge e_3 = 2\alpha_2^+.$$

Similarly

$$J_0 \alpha_2^+ = -2\alpha_1^+, \quad J_0 \alpha_3^+ = 0,$$

$$J_0 \alpha_1^- = J_0 \alpha_2^- = J_0 \alpha_3^- = 0.$$

Thus J_0 has eigenvalues $\pm 2i$, 0 on Λ_+^2, and only 0 on Λ_-^2. Tensored with EndW we see we can only build a 4i eigenvector in the EndW valued self-dual forms. Thus integrability forces $\mathcal{W}_+$ to vanish. A metric with $\mathcal{W}_+ = 0$ is said to be anti-self-dual. Thus we have

Theorem 5 If (M,g) is an oriented Riemannian 4-manifold then J_1 on $J_+(M,g)$ is integrable if and only if the metric is anti-self-dual.

Remarks 1. If we use the bundle $J_-(M,g)$ of orientation disagreeing metric compatible complex structures, then the roles of $\mathcal{W}_+$ and $\mathcal{W}_-$ are reversed, so J_1 is integrable if and only if g is self-dual.

2. $J_+(M,g)$ has several alternative geometrical realisations. It can be identified with the unit sphere bundle in Λ_+^2 and as the projective half spinor bundle $P(V_+)$ when M is also spin. See Atiyah, Hitchin, Singer [1978] or Salamon [1983] for these calculations done from these other points of view.

After orthogonal structures it is natural to look at unitary structures. That is we assume (M,g,J) is an almost Hermitian manifold. The bundle of complex lines in T'M (the 1,0 tangent vectors) can be viewed as the J-stable 2-planes in TM, and given such a 2-plane it determines an endomorphism of the tangent space by reversing J on its orthogonal. This gives us a map

$$j : P(T'M) \longrightarrow J(M,g)$$

which is holomorphic on the fibres and projecting the Levi-Civita connection into the unitary structure gives a connection which can be used to define J_1. The integrability question has been examined by O'Brian and Rawnsley [1985]. They find that for dim M $\geqslant$ 8 we have

<u>Theorem 6</u> J_1 on P(T'M) is integrable if and only if J is integrable on M, g is locally conformal to a Kaehler metric with vanishing Bochner tensor.

In dimension 6 there is an example (S^6 with its almost complex structure coming from the octonions) where J_1 is integrable but the almost complex structure on S^6 is not integrable. In dimension 4 we are back to the self-duality condition since $P(T'M) = J_-(M,g)$ in this case.

The other general family of G-structures is the quaternion structures. Berard-Bergery [1979] and Salamon [1982] have studied a twistor space for these with fibre S^2. The integrability condition is simply to admit a torsion-free connection and for such a connection there is no condition on the curvature.

Finally a large number of examples is furnished by the symmetric spaces G/K whose involution is an inner automorphism of G. If the twistor space is a homogeneous space G/H then J_1 is always integrable (this can be shown directly without using the integrability condition) and G/H is a flag manifold. The structure of these twistor spaces will be described in a forthcoming paper by Burstall and Rawnsley. Special cases where G/H is a 3-symmetric space are described in Bryant [1984] and Salamon [1984].

Lecture V: Self-dual Yang-Mills Fields and Twistors

Let (M,g) be an oriented 4-manifold and Z the bundle $J_-(M,g)$. We saw last time that J_1 defined by the Levi-Civita connection is integrable if the Weyl curvature tensor is self-dual. Now we want to use this twistor space to study the self-dual Yang-Mills connections on M. More precisely, if E now denotes a Hermitian vector bundle over M with connection ∇ then ∇ is a minimum for the Yang-Mills action if its curvature R^∇ is self--dual or anti-self-dual

$$* R^\nabla = \pm R^\nabla.$$

The anti-self-dual solutions can be handled by reversing the orientation to change them into self-dual solutions so we shall just discuss the self-dual case here. The key lemma in the use of the twistor space for Yang-Mills is

Lemma A 2-form δ on M is self-dual if and only if its pull-back to $J_-(M,g)$ is of type 1,1 for J_1.

Proof $\tilde{\delta} = \pi^*\delta$ is type 1,1 if and only if

$$\tilde{\delta}(J_1 X, J_1 Y) = \delta(X,Y) \quad \forall \; X, Y \in TJ_-(M,g).$$

This is equivalent to

$$\delta(\pi_* J_1 X, \pi_* J_1 Y) = \delta(\pi_* X, \pi_* Y)$$

or

$$\delta(JX,JY) = \delta(X,Y) \quad \forall \; J \in J_-(M,g).$$

Fibrewise this means that δ belongs to the largest invariant subspace of $\Lambda^2 W^*$ in the zero eigenspace of J_0 in the Lie algebra representation. To see what this is we use the calculation we did in proving Theorem 4. The only difference is that we are now dealing with $J_-(M,g)$ so Λ_+^2 and Λ_-^2 are interchanged. Thus Λ_+^2 is the only irreducible contained in $\text{Ker} J_0$. Hence the Lemma.

An immediate Corollary to this Lemma is

<u>Corollary</u> ∇ is self-dual if and only if $\pi^{-1}\nabla$ on $\pi^{-1}E$ has curvature of type 1,1 for J_1.

<u>Proof</u> The curvature of a pull-back connection $\pi^{-1}\nabla$ is the pull-back of the curvature of ∇, so the Lemma applies at once.

Consider now the operator

$$(\pi^{-1}\nabla)^{0,1} : \Omega^0(\pi^{-1}E) \longrightarrow \Omega^{0,1}(\pi^{-1}E).$$

We can ask if it is the $\bar{\partial}$-operator of some holomorphic structure in $\pi^{-1}E$. If it is then $\bar{\partial}^2 = 0$ so we shall need

$$0 = ((\pi^{-1}\nabla)^{0,1})^2 = ((\pi^{-1}\nabla)^2)^{0,2} = (\pi^* R^\nabla)^{0,2}.$$

Since ∇ is Hermitian this is equivalent to $(\pi^* R^\nabla)^{2,0} = 0$ and hence that $\pi^* R^\nabla$ is of type 1,1. By the Corollary this is in turn equivalent to R^∇ being self-dual. Since it is known that $\bar{\partial}^2 = 0$ is the only obstruction to putting a holomorphic structure on $\pi^{-1}E$, we see we have obtained the Theorem of Atiyah, Hitchin and Singer [1978]:

<u>Theorem 7</u> R^∇ is self-dual if and only if $(\pi^{-1}\nabla)^{0,1}$ is the $\bar{\partial}$-operator of a holomorphic structure on $\pi^{-1}E$.

In fact $\pi^{-1}\nabla$ and so ∇ can be recovered from the holomorphic structure and the Hermitian structure of $F = \pi^{-1}E$. Which holomorphic bundles on $J_-(M,g)$ arise this way? Being a pull-back, F is trivial on the fibres of $J_-(M,g)$. Further, $J_-(M,g)$ has an involution τ which is antiholomorphic defined by

$$\tau(J) = -J.$$

Such an involution is called a real structure. It preserves the fibres and has no fixed points. Using the Hermitian structure it can be lifted to F to give a holomorphic isomorphism σ of $\tau^{-1}\bar{F}$ with F*. We call such a σ a real structure on F.

In fact this gives a one-one correspondence between self-dual connections on bundles E on M and holomorphic bundles F on $J_-(M,g)$ having a real structure and which are trivial on the fibres. The restriction of the real stucture to the holomorphic sections along any fibre should induce a positive definite Hermitian form. Thus the problem of finding self-dual connections has been translated into studying holomorphic bundles with certain extra structure. It is possible also to encode reductions of the structure group of the original bundle E into the holomorphic structure of F and so obtain self-dual connections for the various compact Lie groups. For example, SU(2)-connections correspond with bundles having a holomorphic skew form on each fibre, and SO(3)-connections to bundles with a holomorphic symmetric bilinear form.

The main example where this has been applied is to $M = S^4$ and we now look at this case in a little more detail. It is convenient to have a concrete realization of $J_-(S^4, can)$. This we obtain by viewing S^4 as the quaternionic projective line $P_1(\mathbb{H})$. The latter can be viewed as the 1-dimensional subspaces of $\mathbb{H}^2$ where $\mathbb{H}$ acts on the right, or equivalently, as the quotient of $\mathbb{H}^2 \setminus \{0\}$ by the action of $\mathbb{H} \setminus \{0\}$

$$(q_1, q_2) \cdot q = (q_1 q, q_2 q).$$

If we view R^5 as $\mathbb{H} \oplus R$ and S^4 as the unit sphere, then

$$(q_1, q_2) \longmapsto \left(\frac{2 q_1 \bar{q}_2}{|q_1|^2 + |q_2|^2}, \frac{|q_1|^2 - |q_2|^2}{|q_1|^2 + |q_2|^2} \right) \in S^4$$

induces an isomorphism of $P_1(\mathbb{H})$ with S^4.

We identify C with the subspace of $\mathbb{H}$ spanned by 1, i and then $\mathbb{H} = \mathbb{C} + j\mathbb{C} = \mathbb{C}^2$ and $\mathbb{H}^2 = \mathbb{C}^4$. If we take the space of complex lines in $\mathbb{C}^4$ we get the complex projective space $P_3(\mathbb{C})$ and taking for any complex line the quaternion line it spans we get a map

$$\pi : P_3(\mathbb{C}) \longrightarrow P_1(\mathbb{H}),$$

a bundle with projective lines as fibres.

If $Q \in P_1(\mathbb{H})$ then the tangent space to $P_1(\mathbb{H})$ at Q can be identified with $\text{Hom}_{\mathbb{H}}(Q, Q^\perp)$ where we calculate the orthogonal with respect to the quaternionic inner product

$$u \cdot v = \bar{u}_1 v_1 + \bar{u}_2 v_2 = u^* v$$

on $\mathbb{H}^2$ where * denotes the conjugate transpose of a quaternion vector or matrix. If we have a complex line $\ell \subset Q$ then we can restrict $A \in \text{Hom}_{\mathbb{H}}(Q, Q^\perp)$ to this line where it is of course complex linear and so is in $\text{Hom}_{\mathbb{C}}(\ell, Q^\perp)$. Since ℓ spans Q over $\mathbb{H}$ it is clear that this restriction determines A. A count of the real dimensions of these spaces shows

$$\text{Hom}_{\mathbb{H}}(Q, Q^\perp) = \text{Hom}_{\mathbb{C}}(\ell, Q^\perp).$$

The right-hand side of this equation is a complex vector space and so the line ℓ in Q has put a complex vector space structure on the tangent space at Q. This gives us a map

$$j : P_3(\mathbb{C}) \longrightarrow J(P_1(\mathbb{H}), \text{can})$$

which makes $P_3(\mathbb{C})$ into a twistor space for $P_1(\mathbb{H})$. In fact it is $J_-(P_1(\mathbb{H}), \text{can})$.

Lemma $P_3(\mathbb{C}) = J_-(P_1(\mathbb{H}), \text{can})$

Proof Because the complex structure corresponds with multiplication by i on $\mathbb{H}$ on the right. If we take the oriented basis

$$e_1 = 1, \quad e_2 = i, \quad e_3 = j, \quad e_4 = k$$

for $\mathbb{H}$ then

$$e_1 i = e_2, \quad e_3 i = -e_4.$$

This clearly has the wrong orientation. Since both bundles have S^2 as fibre the Lemma follows.

<u>Corollary</u> Self-dual connections on bundles on S^4 correspond with holomorphic bundles on $P_3(C)$ having a real structure and which are trivial on the $P_1(C)$ fibres.

The real structure on $P_3(C)$ is easily seen to be the map induced by right multiplication by j on $\mathbb{H}^2$ (since j anticommutes with i).

A theorem of Serre says that all holomorphic bundles on $P_3(C)$ are in fact algebraic. The algebraic bundles have recently been the objects of much study by algebraic geometers, and we can use their results to study the twistor translation of the problem. Horrocks [1964] showed how bundles on $P_3(C)$ are determined by certain linear maps between cohomology groups, and Barth and Hulek [1977] have translated these results into a very simple construction of all the bundles which arise from self-dual connections. Since their work involves sheaf theory and techniques far removed from the differential geometry of this Summer School I shall simply give their answer.

SU(2)-bundles are perhaps the easiest to describe, so I shall limit myself to this case and refer the reader to Atiyah's Pisa lectures [1979] or Manin's book [1985?] for a more comprehensive treatment. An SU(2) connection corresponds with a holomorphic rank 2 bundle on $P_3(C)$ with a real structure and a holomorphic non-degenerate skew-form. There are then vector spaces V, W of dimensions 2k+2, k respectively and a linear map

$$A : \mathbb{C}^4 \longrightarrow \mathrm{Hom}_\mathbb{C}(W,V)$$

such that $A(z)W$ is k dimensional and isotropic for a non-degenerate skew form on V coming from that on the fibres of F. If we set

$$U_z = A(z)W$$

then U_z depends only on the line spanned by z and so gives rise to a bundle U of rank k on $P_3(\mathbb{C})$. Its orthogonal U^o with respect to the skew form will contain U and be a bundle of rank k+2. Thus the quotient U^o/U will be a holomorphic rank 2 bundle on $P_3(\mathbb{C})$, and has a holomorphic skew form induced from that on V. The real structure comes from an antilinear map σ_1 on V preserving the skew form, and a real structure σ_2 on W. A has to satisfy the reality condition

(10) $\qquad \sigma_1 A(z) \sigma_2 = A(zj).$

σ_1 combines with the skew form to produce a positive definite Hermitian structure on V. If F_z denotes the Hermitian orthogonal of U_z in U_z^o, then F is a rank 2 bundle on $P_3(\mathbb{C})$ with a Hermitian structure, and has the holomorphic structure of U^o/U. These Hermitian and holomorphic structures of F combine to produce the Hermitian connection in F which can be obtained directly as follows: Sections of F can be viewed as V-valued functions on $P_3(\mathbb{C})$. Differentiation as functions followed by orthogonal projection back into F gives rise to a connection which can be checked to be both Hermitian and have 0,1 part the $\bar{\partial}$-operator of the holomorphic structure. σ_1 induces the real structure on F and so it follows that F comes from a bundle on $P_1(\mathbb{H})$ and the connection from a self-dual connection. This self-dual connection can be found as follows:

σ_1 acts on V just as j acts on $\mathbb{C}^{2k+2}$ when the latter is identified with $\mathbb{H}^{k+1}$, so we can identify V and $\mathbb{H}^{k+1}$ by choosing a suitable basis for V. If we identify $\mathbb{H}^2$ with $\mathbb{C}^4$ then A(z) becomes an R-linear map

$$P : \mathbb{H}^2 \longrightarrow \text{Hom}_R(R^k, \mathbb{H}^{k+1}).$$

The reality condition (10) tells us that P is in fact quaternion linear, so has the form

$$P(q) = P_1 q_1 + P_2 q_2$$

where P_1, P_2 are quaternionic $k+1 \times k$ matrices. The subspace F_z of V is then identified with the quaternion orthogonal of $P(q)R^k$. The isotropic condition on A can be translated into the condition that $P(q)*P(q)$ which a priori is quaternionic shall actually be real for all q in $\mathbb{H}^2$.

Hence we have produced a quaternionic line bundle and a connection given by orthogonal projection of differentiation. This is thus an $Sp(1) = SU(2)$ connection in an $SU(2)$ bundle on S^4. It is self-dual because we know it comes from a holomorphic bundle on $P_3(\mathbb{C})$.

This process does not necessarily lead to gauge inequivalent connections. In fact if $S \in Sp(k+1)$, $T \in GL(k,R)$ then

$$S \; P(q) \; T$$

will also satisfy the conditions and lead to an equivalent connection. Thus the space of self-dual connections modulo equivalence is the same as the families $P(q)$ of $k+1 \times k$ quaternion matrices for which

$$P(q)*P(q) \in GL(k,R), \qquad q \in \mathbb{H}^2 \setminus \{0\}$$

modulo the above action of $Sp(k+1) \times GL(k,R)$.

The connection can be made quite explicit if we choose an affine coordinate $q = (x,1)$ on $P_1(\mathbb{H})$ (this corresponds with stereographic projection of S^4 onto R^4). We use the action of $Sp(k+1)$ to rotate the range of P_1 to the standard $\mathbb{H}^k \subset \mathbb{H}^{k+1}$, so

$$P_1 = \begin{bmatrix} 1_k \\ 0 \end{bmatrix}.$$

Then we can write

$$P_2 = \begin{bmatrix} A \\ b \end{bmatrix}$$

with A $k \times k$ and b $1 \times k$ and

$$P(q) = \begin{bmatrix} x1+A \\ b \end{bmatrix}.$$

The line in $\mathbb{H}^{k+1}$ orthogonal to $P(q)R^k$ has a section

$$s(x) = (1 + v^*v)^{-\frac{1}{2}} \begin{bmatrix} v \\ 1 \end{bmatrix}$$

where

$$v(x) = -(x1+A)^{*-1}b^*$$

and covariant differentiation is given by projecting ds. This gives the connection form s^*ds which has the form

$$s^*ds = \frac{v^*dv - dv^*v}{2(1 + v^*v)}.$$

This is the general self-dual Yang-Mills potential on S^4.

References

Atiyah,MF. 1979 Geometry of Yang-Mills fields. Lezioni Fermiane, Scuola Normale, Pisa.

Atiyah,MF; Drinfeld,VG; Hitchin,NJ; Manin, YuI. 1978 Construction of instantons. Physics Letters 65A, 185-187.

Atiyah,MF; Hitchin,NJ; Singer,IM. 1978 Self-duality in four-dimensional Riemannian geometry. Proc. Roy. Soc. Lond. A 362 425-461.

Atiyah,MF; Ward,RS. 1977 Instantons and algebraic geometry. Comm. Math. Phys. 55 117-124.

Barth,W; Hulek,K. 1978 Monads and moduli of vector bundles. Manus. Math. 25 323-347.

Berard-Bergery,L. 1979 Quaternionic manifolds. Journees SMF. (Unpublished).

Berard-Bergery,L; Ochiai,T. 1982 On some generalizations of the construction of twistor spaces. Symposium on Global Riemannian Geometry. Horwood-Ellis.

Bryant,R. 1982 Conformal and minimal immersions of compact surfaces into four-manifolds. J. Diff. Geom. 17 455-473.

Bryant,R. 1985 Lie groups and twistor spaces. Duke Math. J. 52 223-261.

Burns,D. 1982 Harmonic mappings from CP^1 CP^n. Lecture Notes in Mathematics 949 48-55. Springer-Verlag, Berlin.

Burstall,FE. 1986 A twistor description of harmonic maps of a two-sphere into a Grassmannian. Math. Ann. 274 61-74.

Burstall,FE; Rawnsley,JH. 1986 Spheres harmoniques dans les groupes de Lie compacts et courbes holomorphes dans les espaces homogenes. C. R. Acad Sci. Paris 302 709-712.

Burstall,FE; Rawnsley,JH. 1987 Twistors harmonic maps and homogeneous geometry. (To appear).

Burstall,FE; Rawnsley,JH; Salamon,S. 1987 Stability of harmonic maps of Riemann surfaces into symmetric spaces. Bull. Amer. Math. Soc. (To appear).

Calabi,E. 1967 Minimal immersions of surfaces in Euclidean spheres. J. Diff. Geom. 1 111-125.

Din,AM; Zakrzewski,WJ. 1980 Properties of the general classical CP^{n-1}-model. Phys. Lett. 95B 419-422.

Dubois-Violette,M. 1983 Structures complexes au-dessus des varietes. Progress in Mathematics vol. 37, Birkhauser, Basel.

Eastwood,MG; Penrose,R; Wells,RO. 1981 Cohomology and massless fields. Commun. Math. Phys. 78 305-351.

Eells,J; Salamon,S. 1983 Constructions twistorielles des applications harmoniques. C. R. Acad. Sci. Paris 296 685-687.

Eels,J; Wood,JC. 1983 Harmonic maps of surfaces into complex projective spaces. Adv. Math. 49 217-263.

Friedrich,Th. 1983 An application of the twistor theory of surfaces in 4-dimensional geometry. Proceedings of the Conference on Differential Geometry, Nove Mesto na Morave, Czechoslovakia.

Friedrich,Th. 1984 On surfaces in four-spaces. Ann. Global. Anal. Geom. 2 257-287.

Glaser,V; Stora,R. 1980 Regular solutions of the CP^n models and further generalizations. (CERN preprint).

Hartshorne,R. 1978 Stable vector bundles and instantons. Commun. Math. Phys. 59 1-15.

Henkin,GM; Manin,YuI. 1980 Twistor description of classical Yang-Mills-Dirac fields. Phys. Lett. 95B 405-408.

Hitchin,NJ. 1979 Polygons and gravitons. Math. Proc. Camb. Phil. Soc. 85 465-476.

Hitchin,NJ. 1980 Linear field equations on self-dual spaces. Proc. Roy. Soc. Lond. A370 173-191.

Hitchin,NJ. 1982 Monopoles and geodesics. Comm. Math. Phys. 83 579-602.

Horrocks,G. 1964 Vector bundles on the punctured spectrum of a local ring. Proc. Lond. Math. Soc. 14 684-713.

Lawson,HB. 1980 Lectures on minimal submanifolds. Publish or Perish.

Manin,YuI. Gauge fields and complex geometry. Plenum Press (To appear).

Murray,MK. 1983 Monopoles and spectral curves. Comm. Math. Phys. 90 263-271.

O'Brian,R; Rawnsley,JH. 1985 Twistor spaces. Ann. Global Anal. Geom. 3 29-85.

Pedersen,H. 1986 Einstein metrics, spinning top motions and monopoles. Math. Ann. 274 35-59.

Penkov,IB. 1980 The Penrose transform on general Grassmannians. C. R. Acad. Sci. Bulg. 33 1439-1442.

Penrose,R. 1967 Twistor algebra. J. Math. Phys. 8 345-366.

Penrose,R. 1977 The twistor programme. Rep on Math. Phys. 12 65-76.

Poon,YS. 1985 Self-duality. Thesis,Oxford.

Rawnsley,JH. 1985 f-structures, f-twistor spaces and harmonic maps. Seminar 'Luigi Bianchi', Lecture Notes in Mathematics 1164, Springer-Verlag, Berlin.

Rawnsley,JH. 1987 Stability of harmonic maps into symmetric spaces. Proceedings of the Colloque applications harmoniques, CIRM, Luminy, 1986. Travaux en cours (to appear).

Robinson,I. 1961 Null electromagnetic fields. J. Math. Phys. 2 290-291.

Salamon,S. 1980 Quaternionic manifolds. Thesis, Oxford.

Salamon,S. 1983 Topics in four-dimensional Riemannian geometry. Seminar 'Luigi Bianchi', Lecture Notes in Mathematics 1022, Springer-Verlag, Berlin.

Salamon,S. 1985 Harmonic and holomorphic maps. Seminar 'Luigi Bianchi', Lecture Notes in Mathematics 1164, Springer-Verlag, Berlin.

Uhlenbeck,K. 1985 Harmonic maps into Lie groups. J. Diff. Geom. (to appear).

Veblen,O. 1933 Geometry of four-component spinors. Proc. Nat. Acad Sci. USA 19 503-517.

Weierstrass,K. 1866 Monatsberichte der Berliner Akademie. 612-625.

Wells,RO. 1979 Complex manifolds and mathematical physics. Bull. Amer. Math. Soc. 1 296-336.

Witten,E. 1978 An interpretation of classical Yang-Mills theory. Phys. Lett. 77B 394-398.

Partial Differential Equations

in Differential Geometry

Jerry L. Kazdan[*]

These lectures will first treat some of the basic facts about linear elliptic equations and then investigate a few nonlinear problems. The goal is to give a rapid intuitive introduction to advanced graduate students. Our main emphasis will be to understand some of the basic examples which are the backbone of the subject. Almost no proofs are given.

Lecture I

1. Classical Equations

Three models from classical physics are the source of most of our knowledge of partial differential equations:

$u_{tt} = u_{xx} + u_{yy}$ wave equation

$u_t = u_{xx} + u_{yy}$ heat equation

$0 = u_{xx} + u_{yy}$ Laplace equation

One thinks of a solution $u(x,y,t)$ of the wave equation as describing the motion of a drumhead Ω at the point (x,y) at time t. A typical problem is to specify

initial position $u(x,y,0)$, initial velocity $u_t(x,y,0)$
boundary conditions $u(x,y,t)$ for $(x,y) \in \partial\Omega$, $t \geq 0$

and seek the solution $u(x,y,t)$.

For the heat equation, $u(x,y,t)$ represents the temperature at (x,y) at time t. Here a typical problem is to specify

---

[*] Research supported in part by a grant from the National Science Foundation.

initial temperature $\quad u(x,y,0)$

boundary temperature $\quad u(x,y,t)$ for $(x,y) \in \partial\Omega$, $t \geq 0$

and seek $u(x,y,t)$ for $(x,y) \in \Omega$, $t > 0$.

It is clear that if a solution $u(x,y,t)$ is <u>independent</u> of t, so one is in equilibrium, then u is a solution of the <u>Laplace equation</u> (these are called <u>harmonic functions</u>). Using the heat equation model, a typical problem is the <u>Dirichlet problem</u>, where one is given

boundary temperature $u(x,y)$ for $(x,y) \in \partial\Omega$

and one seeks the (equilibrium) temperature distribution $u(x,y)$ for $(x,y) \in \Omega$. From this physical model, it is intuitively evident that in equilibrium, the maximum (and minimum) temperatures can not occur at an interior point of Ω unless $u \equiv$ const., for if there were a local maximum temperature at an interior point of Ω, then the heat would flow away from that point and contradict the assumed equilibrium. This is the <u>maximum principle</u>: if u satisfies the Laplace equation then

$$\min_{\partial\Omega} u \leq u(x,y) \leq \max_{\partial\Omega} u \qquad \text{for } (x,y) \in \Omega.$$

Of course, one must give a genuine mathematical proof as a check that the differential equation really does embody the qualitative properties predicted by physical reasoning such as this.

For many mathematicians, a more familiar occurrence of harmonic functions is as the real (or imaginary) parts of analytic functions. Indeed, one should expect that harmonic functions have all of the properties of analytic functions - with the important exception that the product of two harmonic functions is almost never harmonic.

2. An Example

In elementary courses in differential equations one main task is to find explicit formulas for solutions of differential equations. This can only be done in the simplest situations, the resulting formulas being fundamental in more advanced work where one must gain insight without such explicit formulas. As a typical situation, let (M^n, g) be a compact Riemannian manifold without boundary, for instance the torus T^2 with its flat metric (so the functions on T^2 are doubly periodic function on the plane, $\mathbb{R}^2$). We wish to solve the heat equation

(1) $\quad u_t = \Delta u$, $\qquad\qquad x \in M$

where Δ is the Laplacian of the metric g, which in local coordinates $x = (x_1,\ldots,x_n)$ is

$$\Delta u = \frac{1}{\sqrt{g}} \sum_{i,j=1}^{n} \frac{\partial}{\partial x^i}\left(g^{ij}\sqrt{g}\,\frac{\partial u}{\partial x^j}\right)$$

with $\sqrt{g} = \sqrt{\det g}$ and g^{ij} the inverse of the metric g_{ij} (for the flat torus, $g_{ij} = \delta_{ij}$ of course). Our initial condition is

(2) $\quad u(x,0) = f(x)$,

where f is a prescribed function on M.

Guided by ordinary differential equations we can write the "solution" as

(3) $\quad u(x,t) = e^{t\Delta} f$.

To make sense of this we use a spectral representation of Δ. Thus, let λ_j and ϕ_j be the eigenvalues and corresponding eigenfunctions of $-\Delta$

(4) $\quad -\Delta\phi_j = \lambda_j \phi_j$.

By general theory, for any (M,g) the λ_j's are a discrete set of real numbers converging to ∞ and a complete (in $L_2(M)$) set of orthonormal eigenfunctions. Moreover, multiplying (4) by ϕ_j and integrating by parts (= the divergence theorem)

(4)' $\quad \lambda_j = \dfrac{\int |\nabla \phi_j|^2 dx}{\int \phi_j^2 dx} \geq 0$,

where dx is the Riemannian element of volume on (M,g). Formally, we seek a solution as an eigenfunction expansion

$$u(x,t) = \sum a_j(t)\phi_j(x).$$

Substituting this into (1) and using the initial condition we obtain

(5) $\quad u(x,t) = \sum_j f_j\, e^{-\lambda_j t}\, \phi_j(x)$,

where

$$f_j = \int f(y)\phi_j(y)\, dy.$$

One can rewrite this solution (5) as

(6) $\quad u(x,t) = \int H(x,y;t)f(y) \, dy$,

with

(7) $\quad H(x,y;t) = \sum_j e^{-\lambda_j t} \phi_j(x)\phi_j(y)$.

This function H is called the <u>heat kernel</u> or <u>Green's function</u> for the problem (1) - (2). The formulas (6) - (7) is our interpretation of (3), so $e^{t\Delta}$ is an integral operator (6) with kernel H. Then

(8) $\quad \text{trace } e^{t\Delta} = \int H(y,y;t) \, dy = \sum_j e^{-\lambda_j t}$.

We will use this formula later. Of course, it is very difficult to extract much information from (6) - (7) unless one has more information on the λ_j's, ϕ_j's or some formula other than (7) giving properties of H. These properties depend on the manifold M as well as the metric g.

One simple consequence is

(9) $\quad \lim_{t \to \infty} u(x,t) = \text{average of } f = \frac{1}{\text{Vol}(M)} \int f \, dx$.

To prove this, one notes from (4)' that $\lambda_0 = 0$, $\lambda_j > 0$ for $j \geq 1$ and $\phi_0(x) = \text{constant} = \text{Vol}(M)^{-\frac{1}{2}}$. Then by (7)

$$\lim_{t \to \infty} H(x,y,t) = \text{Vol}(M)^{-1}$$

so the assertion now follows from (6). The formula (9) states that the equilibrium temperature is the average of the initial temperature - which is hardly surprising.

3. Hölder and Sobolev Spaces

From calculus one knows that

regularity: if $u'' = f \in C^k$ then $u \in C^{k+2}$

existence : given any $f \in C^k$ then $u'' = f$ for some $u \in C^{k+2}$.

Thus, one might anticipate that, at least locally,

(10a) $\quad$ if $\Delta u = f \in C^k$ then $u \in C^{k+2}$

and

(10b)   given any $f \in C^k$ then $\Delta u = f$ for some $u \in C^{k+2}$.

Both of these last two assertions are FALSE except in dimension one. But they are almost true. The trouble is that the spaces C^k are not really appropriate. After a century we have learned to use the Hölder spaces $C^{k+\alpha}$, where $0 < \alpha < 1$, and Sobolev spaces H_k^p, $1 < p < \infty$ (here the p is as in the Lebesgue spaces L^p). If in (10a,b) one replaces C^k and C^{k+2} by $C^{k+\alpha}$ and $C^{k+2+\alpha}$ (or by H_k^p and H_{k+2}^p), then the assertions become true.

With this as motivation, we define these spaces. Let $\Omega \subset \mathbb{R}^n$ be an open set. Then for $0 < \alpha < 1$,

$$\|f\|_{C^\alpha(\bar{\Omega})} = \sup_{\substack{x,y \in \Omega \\ x \neq y}} \frac{|f(x)-f(y)|}{|x-y|^\alpha}$$

and the Hölder space $C^{k+\alpha}(\bar{\Omega})$ has the norm

$$\|f\|_{C^{k+\alpha}(\bar{\Omega})} = \|f\|_{C^k(\bar{\Omega})} + \max_{|j|=k} \|\partial^j f\|_{C^\alpha(\bar{\Omega})},$$

where ∂^j with $|j| = k$ refers to any partial derivative of order k. The standard example of a function $f \in C^\alpha$ is $f(x) = |x|^\alpha$ near the origin. On a manifold, one defines these spaces using a partition of unity. These spaces are all Banach spaces.

The Sobolev space $H_k^p(\Omega)$ is defined as the functions f whose derivatives up to order k are all in L^p, with the norm

$$\|f\|_{H_k^p(\Omega)} = \left(\int \sum_{|j| \leq k} |\partial^j f|^p \, dx \right)^{1/p}, \quad 1 \leq p < \infty.$$

An equivalent definition is as the completion of $C^\infty(\Omega)$ in the $H_k^p(\Omega)$ norm. In what follows, we always assume $1 < p < \infty$ (i.e. exclude the more awkward case $p = 1$). The special case where $p = 2$ gives Hilbert spaces with the obvious inner products. For <u>linear</u> partial differential equations with C^∞ coefficients, the Hilbert space case $p = 2$ is often adequate. However, <u>nonlinear</u> equations force one to use all $1 < p < \infty$ (and even the case $p = 1$, which we ignore). Note that various other notations are used:

$$H_k^p, \quad H_{p,k}, \quad L_k^p, \quad W_{p,k}$$

for these same spaces. The special case $p = 2$ is often simply written H_k.

It is important to relate all of these spaces to each other and to the familiar spaces $C^k(\Omega)$. This is quite simple if $\Omega = \{0 < x < 1\}$ in $\mathbb{R}^1$ since

$$u(x) = u(y) + \int_y^x u' \le |u(y)| + \int_0^1 |u'|$$

so, integrating this with respect to y we obtain

$$|u(x)| \le \int_0^1 (|u'| + |u|),$$

that is,

$$\|u\|_{C^0} \le \|u\|_{H_1^1} \le \|u\|_{H_1^p} \qquad \text{for any } p \ge 1$$

(the last inequality is a consequence of Hölder's inequality). Thus, a Cauchy sequence in H_1^p is also Cauchy in C^0, so we have a continuous embedding of $H_1^p \hookrightarrow C^0$.

In higher dimensions, $\Omega \subset \mathbb{R}^n$, the story is similar but more complicated. The result is called the <u>Sobolev embedding theorem</u>. Say $f \in H_1^p(\Omega)$,

(12)   (a) If $1 < \frac{n}{p}$ and $\frac{1}{p} - \frac{1}{n} \le \frac{1}{q}$, then $f \in L^q(\Omega)$.

      Also

(13) $\|f\|_{L^q(\Omega)} \le c \|f\|_{H_1^p(\Omega)}$

where the constant c does not depend on f. Thus there is a continuous embedding $H_1^p(\Omega) \hookrightarrow L^q(\Omega)$.

(14)   (b) If $\frac{n}{p} < 1$ then $f \in C^\alpha(\Omega)$, where $\alpha = 1 - \frac{n}{p}$.

      Also

$$\|f\|_{C^\alpha(\Omega)} \le c \|f\|_{H_1^p(\Omega)}.$$

Thus there is a continuous embedding $H_1^p(\Omega) \hookrightarrow C^\alpha(\Omega)$.

  (c) Moreover, if Ω is bounded with smooth boundary and, for (a), if the strict inequality in (12)

(15) $\frac{1}{p} - \frac{1}{n} < \frac{1}{q}$,

holds, then these embeddings are compact, i.e. the injections $H_1^p \hookrightarrow L^q$ and $H_1^p \hookrightarrow C^\alpha$ are compact linear maps.

Consequences:

i) $H^p_{k+1} \hookrightarrow H^p_k$, and is compact if Ω is smoothly bounded

ii) $H^p_2 \hookrightarrow C^0$ if $2p > n$, and $H^p_k \hookrightarrow C^0$ if $kp > n$

iii) $C^\infty = \cap_k H^p_k$

The first, (i), follows from part (a). For (ii), if $2p > n$, then by part (a) $H^p_2 \hookrightarrow H^q_1$ for $\frac{1}{q} = \frac{1}{p} - \frac{1}{n} < \frac{1}{n}$ so the result follows from (b). Repeating this one has $H^p_k \hookrightarrow C^0$ if $kp > n$. To prove (iii), use (ii) to find that $H^p_{k+1} \hookrightarrow C^1$ if $kp > n$, etc.

Note that the restrictions (12), (14) are easy to find. Say inequality (13) holds in a ball $\{|x|<1\} \subset \mathbb{R}^n$. Pick some $\phi \in C^\infty_0(|x|<1)$, $\phi \not\equiv 0$ and apply (13) to $f_\lambda(x) = \phi(\lambda x) \in C^\infty(|x|<1)$ for $\lambda > 1$. After a change of variables in the integrals in (13) one has

$$\|\phi\|_{L^q} \leq c \, \lambda^{\text{power}} \|\phi\|_{H^p_1}$$

where power $= 1 + n(1/q - 1/p)$. This must hold for all $\lambda > 1$. Letting $x \to \infty$ we obtain the contradiction $\|\phi\|_{L^q} = 0$ unless power ≥ 0, which is precisely the condition (12). In this reasoning we use the change of scale = conformal map $x \mapsto \lambda x$ to find the critical case of the Sobolev embedding theorem where equlity hold in (15). One consequence is that in any geometric problem involving conformal invariance one should expect to be in the critical case of the Sobolev embedding theorem. This is precisely what one finds in the Yamabe problem (see Lecture V) and in four dimensional Yang-Mills fields (see the other lectures in this volume).

Lecture II

1. Linear Algebra

Let L be a matrix, not necessarily square. When can one solve the simultaneous linear equations $Lx = y$? One useful way to answer this is to use the adjoint matrix, L^*. Then one can solve $Lx = y$ if and only if y is orthogonal to the kernel of L^*. More abstractly, if V and W are finite dimensional linear spaces with inner products and $L: V \to W$, then there is an orthogonal decomposition

(1) $W = \text{Im}(L) \oplus \ker L^*$

(proof: $Z \perp \text{Im } L \iff$ for all x one has $0 = \langle Lx, z \rangle = \langle x, L^*z \rangle \iff L^*z = 0$).
Thus, the cokernel of L is just $\ker L^*$. In addition one can define the <u>index</u> of L as

(2) $i(L) = \dim \ker L - \dim \ker L^*$

and can prove easily that

(2)' $i(L) = \dim V - \dim W$

<u>independent</u> of L. This becomes more interesting later on.

The basic decomposition (1) is also true for elliptic differential operators. This profound observation is a consequence of the insight of Fredholm at the turn of the century. To state these results, we define the adjoint of a differential operator and define an elliptic differential operator.

2. The Adjoint

On $\mathbb{R}^1$ with the L^2 inner product, the adjoint of $D = d/dx$ is found simply by integrating by parts: for all $\phi, \psi \in C_0^\infty$ (i.e. compact support)

$$\langle \phi, D\psi \rangle = \int \phi \bar{\psi}' \, dx = -\int \phi' \bar{\psi} \, dx = \langle -D\phi, \psi \rangle .$$

Thus, the adjoint of d/dx is $-d/dx$. Actually, this is the <u>formal adjoint</u> since the strict Hilbert space adjoint requires additional attention to the domain of definition of the operator because d/dx is an unbounded operator on L^2.

For smooth Hermitian vector bundles E,F over a Riemannian manifold (M,g) and a differential operator $L: C^\infty(E) \to C^\infty(F)$, one defines the formal adjoint by the usual rule

$$\langle Lu,v \rangle_F = \langle u, L^*v \rangle_E$$

for all smooth sections $u \in C_0^\infty(E)$, $v \in C_0^\infty(F)$ with compact support. Since the supports of u and v can be assumed to lie in a coordinate patch, then one can compute L* locally using integration by parts. Thus, for the Laplacian one has

$$\langle \Delta u, v \rangle = \int v \Delta u = \int u \Delta v = \langle u, \Delta v \rangle$$

so $\Delta^* = \Delta$, that is, the Laplacian is formally self-adjoint. (This basic observation was emphasized by Green early in the last century in what we now call "Green's second identity").

3. The Symbol

It is often the case that the highest order terms in a differential operator are the most important ones. This is formalized by introducing the principal symbol of the operator. If in local coordinates

$$Lu = \sum_{|j| \leq k} a_j(x) \partial^j u ,$$

then the principal symbol of L at x is

(3) $$\sigma_\xi(L;x) = \sum_{|j|=k} a_j(x) \xi^j ,$$

where $j = (j_1, \ldots, j_n)$ is a multi-index and $\xi^j = \xi_1^{j_1} \ldots \xi_n^{j_n}$. (Note that sometimes it is convenient to define the symbol (3) using an additional factor of $(\sqrt{-1})^k$). If the a_j are matrices, then for each x the symbol is a matrix whose elements are polynomials in ξ, homogeneous of degree k.

The symbol can be defined invariantly. Let E_x and F_x be the fibres of E and F at $x \in M$, let $u \in C^\infty(E)$ with $u(x) = z$ and let $\phi \in C^\infty(M)$ have $\phi(x) = 0$, $d\phi(x) = \xi$. Then for each cotangent vector ξ at x the principal symbol of a linear differential operator $L: C^\infty(E) \to C^\infty(F)$ of order k is the following endomorphism $\sigma_\xi(L;x): E_x \to F_x$

(4) $$\sigma_\xi(L;x)z = \frac{1}{k!} L(\phi^k u)\Big|_x .$$

Since the symbol is homogeneous of degree k in ξ, one often just uses unit cotangent vectors.

As a simple example we compute the symbol of the exterior derivative on p-forms, d: $\Omega^p(M) \to \Omega^{p+1}(M)$. This is a first order operator and satisfies $d(\phi\alpha) = d\phi \wedge \alpha + \phi d\alpha$ for any $\phi \in C^\infty(M)$ and any $\alpha \in \Omega^p(M)$. By the rule (4) we find

(5) $\sigma_\xi(d;x)\alpha = \xi \wedge \alpha$,

so the symbol is just exterior multiplication by ξ. Similarly, for any vector bundle E over M, given a section $v: M \to E$ the covariant derivative $Dv: M \to T^*M \otimes E$ has the property $D(\phi v) = d\phi \otimes v + \phi Dv$. Consequently, $\sigma_\xi(D)v = \xi \otimes v$.

4. Linear Elliptic Operators

Many of the properties of the Laplacian and the Cauchy-Riemann equations are shared by a larger class of differential operators, those we call __elliptic__. An operator $L: C^\infty(E) \to C^\infty(F)$ (between smooth sections of vector bundles E and F over M) is elliptic at the point $x \in M$ if the symbol $\sigma_\xi(L;x): E_x \to F_x$ is an isomorphism for all real $\xi \neq 0$. A necessary condition is clearly that $\dim E_x = \dim F_x$. Also, if L is elliptic, so is L^*.

Example 1 The scalar operator with real coefficients

$$Lu = \sum a_{ij}(x) \frac{\partial^2 \phi}{\partial x_i \partial x_j} + \sum b_j(x) \frac{\partial \phi}{\partial x_j} + c(x)\phi$$

has as its principal symbol the 1×1 matrix

$$\sigma_\xi(L;x) = \sum a_{ij}(x)\xi_i \xi_j \ .$$

Thus, L is elliptic at x if the matrix $a_{ij}(x)$ is positive or negative definite at x. For instance, the Laplace equation is elliptic everywhere, the wave and heat equations are not elliptic, while $u_{xx} + xu_{yy} = f(x,y)$ is elliptic only where $x > 0$.

Example 2 The __Cauchy-Riemann operator__ on $\mathbb{R}^2$,

$$L \begin{pmatrix} u \\ v \end{pmatrix} = \begin{pmatrix} u_x & -v_y \\ u_y & v_x \end{pmatrix} ,$$

has symbol

$$\sigma_\xi(L) = \begin{pmatrix} \xi_1 & -\xi_2 \\ \xi_2 & \xi_1 \end{pmatrix}$$

which is an isomorphism so L is elliptic everywhere. We can also write L, using complex coefficients, acting on complex-valued functions f as $Lf = \frac{1}{2}(f_x + i f_y)$. Then $\sigma_\xi(L) = \frac{1}{2}(\xi_1 + i\xi_2)$ is a 1×1 complex matrix, which is clearly invertible if $\xi \neq 0$. Similarly, we can use any pair of real vector fields in $\mathbb{R}^2$

$$L_1 = a_1 \partial/\partial x + b_1 \partial/\partial y \quad , \quad L_2 = a_2 \partial/\partial x + b_2 \partial/\partial y$$

to define an operator $Lf = (L_1 + iL_2)f$. This operator L is elliptic at every point where the vector fields are linearly independent.

Example 3 For the <u>de Rham complex</u> of an n-dimensional Riemannian manifold,

$$\Omega^0(M) \underset{d^*}{\overset{d}{\rightleftarrows}} \Omega^1(M) \underset{d^*}{\rightleftarrows} \cdots \underset{d^*}{\rightleftarrows} \Omega^n(M)$$

one can define the Hodge Laplacian $\Delta_H: \Omega^p(M) \to \Omega^p(M)$ by

(6) $\quad \Delta_H = -(dd^* + d^*d)$.

Let A_ξ denote exterior multiplication by ξ, so from (5), $\sigma_\xi(d) = A_\xi$. Then, using an orthonormal basis with $\xi = \lambda e_1$, one can verify that Δ_H is elliptic

$$\sigma_\xi(\Delta_H) = -(A_\xi A_\xi^* + A_\xi^* A_\xi) = -|\xi|^2 I \ .$$

More generally, given any first order operators P,Q between section of Hermitian vector bundles E,F,G (over M)

$$C^\infty(E) \underset{P^*}{\overset{P}{\rightleftarrows}} C^\infty(F) \underset{Q^*}{\overset{Q}{\rightleftarrows}} C^\infty(G) \ ,$$

one can define

$$L = -(PP^* + Q^*Q): C^\infty(F) \to C^\infty(F) \ .$$

Then L is elliptic at x if the sequence

$$E_x \overset{\sigma_\xi(P;x)}{\to} F_x \overset{\sigma_\xi(Q;x)}{\to} G_x$$

is exact for all real $\xi \neq 0$.

5. Existence for Linear Elliptic Equations

For simplicity we restrict our discussion to compact Riemannian manifolds (M,g) <u>without</u> boundary. If there is a boundary, there are similar results but one must add appropriate boundary conditions - while if (M,g) is not compact, then very little is known except in very special circumstances. Let

$$L: C^\infty(E) \to C^\infty(F)$$

be an elliptic operator of order k. When can one solve Lu = f? The main result is simply the decomposition (1) again

(7) $\qquad C^\infty(F) = \text{Im } L(C^\infty(E)) \oplus \ker L^*$

coupled with the additional pleasant facts that ker L and ker L* are both finite dimensional. One way to prove (7) is to use the Sobolev spaces H_k^2 and prove the orthogonal decomposition

(7)' $\qquad H_{k+\ell}^2(F) = \text{Im} L(H_\ell^2(E)) \oplus \ker L^*$

and then let $\ell \to \infty$. A key fact is that the kernels of L and L* are smooth, in fact, if L is any kth order linear elliptic operator with C^∞ coefficients and $u \in H_k$ satisfies Lu = 0, then $u \in C^\infty$ (this assertion remains true even for "distribution solutions" u). Because of this, as one changes ℓ in (7)', the space ker L* does not change.

<u>Example 0</u> Let Lu = u' + a(x)u for x on the circle S^1 and $a \in C^\infty(S^1)$ a real valued function. Then L*v = - v' + a(x)v, dim ker L* ≦ 1 and one can solve u' + a(x)u = f(x) $\in C^\infty(S^1)$ if and only if $\int_{S^1} fz = 0$ for all $z \in \ker L^*$; the special case where a(x) ≡ 0 so z(x) ≡ 1 is especially obvious, but it is (very) elementary to give a direct proof of the existence assertion for any coefficient a(x) using explicit formulas.

<u>Example 1</u> Consider Lu = - Δu + c(x)u = f on (M,g). Then L = L* (presuming c(x) is real). If c(x) > 0 then ker L = 0 because if Lz = 0, then integrating by parts we find

$$0 = \langle z, Lz \rangle = \int z(-\Delta z + cz) = \int |\nabla z|^2 + cz^2$$

so $z \equiv 0$. Thus, if $c > 0$ then for any $f \in C^\infty$ there is a unique solution $u \in C^\infty$ of $Lu = f$. If $c \equiv 0$ then clearly $\ker L = \ker L^* =$ the constants, so one can solve $-\Delta u = f$ if and only if $\int f = 0$. (In the special case of the flat torus, T^n, one can give short direct proof using Fourier series.)

Example 2 Since the Hodge Laplacian $\Delta_H: \Omega^p(M) \to \Omega^p(M)$ (see (6)) is elliptic, the decomposition (7) gives

$$\Omega^p(M) = \text{Im } \Delta_H(\Omega^p) \oplus \ker \Delta_H .$$

In other words, given any $\omega \in \Omega^p(M)$ there is a $u \in \Omega^p$ and a harmonic p-form h so that

$$\omega = -\Delta_H u + h$$

that is,

(8) $\qquad \omega = d\alpha + d^*\beta + h$, $\qquad\qquad$ with $\alpha = d^*u$, $\beta = du$.

Thus, one has the more refined decomposition

(8)' $\qquad \Omega^p = d(\Omega^{p-1}) \oplus d^*(\Omega^{p+1}) \oplus \mathcal{H}^p$,

where $\mathcal{H}^p$ is the space of harmonic p-forms. The three spaces on the right side of (8)' are easily seen to be orthogonal in the L^2 inner product; for instance, if $\alpha \in \Omega^{p-1}$, $\beta \in \Omega^{p+1}$, then since $d^2 = 0$

$$\langle d\alpha, d^*\beta \rangle = \langle d^2\alpha, \beta \rangle = 0 .$$

Note also that $\Delta_H h = 0$ if and only if h is both closed and co-closed because

$$0 = \langle h, -\Delta_H h \rangle = \langle h, dd^*h \rangle + \langle h, d^*dh \rangle = \|d^*h\|^2 + \|dh\|^2 .$$

As a special case of (8), if ω is closed, then

$$0 = d\omega = d^2\alpha + dd^*\beta + dh = dd^*\beta .$$

But this implies $d^*\beta = 0$ since $0 = \langle \beta, dd^*\beta \rangle = \langle d^*\beta, d^*\beta \rangle$. Consequently, we can write any closed ω as $\omega = d\alpha + h$. Now the de Rham cohomology class $H^p_{deR} = \{\text{closed p-forms/exact p-forms}\}$, so we find that we can represent each de Rham cohomology class by a unique harmonic form. Therefore H^p_{deR} is isomorphic to the space of harmonic p-forms. But by de Rham's theorem, the dimension of $H^p_{deR} = \beta_p =$ pth Betti number, which implies

(9) $\dim \mathcal{H}^p = \beta_p$.

One can use this to extract topological information by studying harmonic forms since they uniquely represent cohomology classes.

The basic decomposition (7)' is a consequence of fundamental inequalities for elliptic operators. For a kth order linear elliptic operator $L: C^\infty(E) \to C^\infty(F)$ there are constants $c_1, \ldots, c_6$ so that

(a) (<u>Schauder estimates</u>) for every $u \in C^{k+\ell+\alpha}(E)$

(10) $\|u\|_{C^{k+\ell+\alpha}} \leq c_1 \|Lu\|_{C^{\ell+\alpha}} + c_2 \|u\|_{C^0} \leq c_3 \|u\|_{C^{k+\ell+\alpha}}$

(b) ($\underline{L^p}$ <u>estimates</u>) for every $u \in H^p_{k+\ell}(E)$

(11) $\|u\|_{H^p_{k+\ell}} \leq c_4 \|Lu\|_{H^p_\ell} + c_5 \|u\|_{L^2} \leq c_6 \|u\|_{H^p_{k+\ell}}$

Moreover, if u is orthogonal in $L^2(E)$ to ker L, then we can let $c_2 = c_5 = 0$. (Note that the inequalities on the right in (10) and (11) are obvious). The moral of (10) - (11) is that in these Hölder and Sobolev spaces, $\|Lu\|$ defines a norm equivalent to the standard norms - except that one must add an extra term if ker L ≠ 0 since then $\|Lu\|$ is only a semi-norm.

To see how to obtain the decomposition (7)' with $\ell = 0$ from (11) we observe that $L: H_k \to L_2$ is continuous so the proof of (1) gives ker $L^* = (\text{Im } L(H_k))^\perp$. Therefore, to prove (7)' with $\ell = 0$ it is enough to show that Im $L(H_k)$ is a closed subspace (in any Hilbert space, $(V^\perp)^\perp = \overline{V}$). Because the injection $H_k \hookrightarrow L_2$ is compact, one need just prove the following Lemma from functional analysis which was observed by Peetre.

<u>Lemma</u> Let X, Y, and Z be reflexive Banach spaces with X $\hookrightarrow$ Y a compact injection and $L: X \to Z$ a continuous linear map. Then the following are equivalent:

a) the image L(X) is closed and ker L is finite dimensional,

b) there are constants c_1 and c_2 such that for all $x \in X$

(12) $\|x\|_X \leq c_1 \|Lx\|_Z + c_2 \|x\|_Y$.

To prove a) $\Rightarrow$ b) write $X = X_1 \oplus \ker L$ so the restriction of L to X_1 is injective and one can appeal to the closed graph theorem to obtain (12).

To prove b) $\Rightarrow$ a), since $X \hookrightarrow Y$ is compact, one sees that the unit ball in ker L is compact so ker L is finite dimensional. Now decompose $X = X_1 \oplus \ker L$. Because $L: X_1 \to Z$ is injective and $X \hookrightarrow Y$ is compact, reasoning by contradiction one finds that all $x \in X_1$ satisfy

(13) $\quad \|x\|_X \leq c\|Lx\|_Z$

with some (new) constant c. Say $Lx_j \to z$ for some $x_j \in X_1$. To show that ImL is closed we find x in X, so that $z = Lx$. But (13) implies the x_j are Cauchy in X so $x_j \to x$ for some $x \in X_1$. Now by continuity $z = \lim Lx_j = Lx$.

Lecture III

1. Bochner's Vanishing Theorem

In the last lecture, we wrote the Hodge Laplacian only in an abstract form (6). It can also be expressed in a different way which is often useful:

(1) $-\Delta_H u = \nabla^* \nabla u + \text{(curvature)} \, u$,

where ∇ is the covariant derivative and "curvature" stands for an expression involving the curvature of the manifold. Formulas like (1) are often called **Weitzenböck formulas**. The curvature expression in (1) is quite simple in the special case of 1-forms when it becomes

(2) $-\Delta_H u = \nabla^* \nabla u + \text{Ric} \, u^\#$,

where u is a 1-form, $u^\#$ is the dual vector field (found using the Riemannian metric g) and Ric is the Ricci curvature of g. If one multiplies (2) by u and integrates by parts, one obtains

$$\langle u, -\Delta_H u \rangle = \int \left[|\nabla u|^2 + \text{Ric}(u^\#, u^\#) \right].$$

If one assumes that Ric > 0, it is clear that the only harmonic 1-form, $-\Delta_H u = 0$, is $u = 0$. But we know that the dimension of the space of harmonic 1-forms is the first Betti number (see II (9)). Consequently, if a compact manifold has $\beta_1 \neq 0$ then there is no Riemannian metric with positive Ricci curvature. This is **Bochner's vanishing theorem**. The identical procedure can be used in other situations, and requires two ingredients:

i) a Weitzenböck-type formula (1) where the "curvature" term is interesting, and

ii) some topological interpretation of the kernel of the operator.

2. Index of an Elliptic Operator

We defined the <u>index</u> $i(L)$ of a linear map $L: \mathbb{R}^n \to \mathbb{R}^k$ and found that $i(L) = n - k$ independent of L (see II (2)). For a linear elliptic operator $L: C^\infty(E) \to C^\infty(F)$ both ker L and ker L^* are finite dimensional so the index still makes sense, and can be defined for any abstract <u>Fredholm operator</u> (i.e. those with finite dimensional kernel and cokernel). The abstract theory states that $i(L)$ depends continuously on L - and hence is constant under continuous change of L, and also that it does not change if one adds a compact operator to L.

Now for a linear elliptic operator L: $H_m^2 \to L^2$ of order m, we can write L as $L = L_m + Q$, where L_m involves only derivatives of order m while Q contains all the lower order derivatives. Because $Q: H_{m-1}^2 \to L^2$ is continuous and $H_m^2 \hookrightarrow H_{m-1}^2$ is compact by the Sobolev theorem, we find that $Q: H_m^2 \hookrightarrow H_{m-1}^2 \to L^2$ is a compact perturbation of L. Consequently, the index of L depends only on the highest order terms, so all the information on the index of L is contained in the symbol of L. Around 1958, I. Gelfand pointed out that one should compute the index in terms of topological data. Atiyah and Singer obtained the general formula in 1963. The result has been enormously powerful and useful.

3. An Example

As a simple example we consider the de Rham complex. Now Δ_H is self-adjoint, so its index is zero. To obtain something more interesting, let $\Omega = \bigoplus_p \Omega^p$ be the algebra of all differential forms under exterior multiplication and let

(3) $D = d + d^* : \Omega \to \Omega$.

Then $D^2 = DD^* = dd^* + d^*d = -\Delta_H I$ so D is elliptic - and is a square root of the Laplacian. We can split Ω into its even and odd parts

(4) $\Omega^{ev} = \bigoplus_{p \text{ even}} \Omega^p$, $\Omega^{odd} = \bigoplus_{p \text{ odd}} \Omega^p$

and let $D^+: \Omega^{ev} \to \Omega^{odd}$ be the restriction of D to Ω^{ev}.

Similarly, we define $D^-: \Omega^{odd} \to \Omega^{ev}$ and note $D^- = (D^+)^*$. D^+ is elliptic (because D is). Also ker D = ker D^2 (if $D^2 u = 0$ then $0 = \langle u, D^2 u \rangle = \|Du\|^2$ so Du = 0). If we let $\mathcal{H}^p$ be the space of harmonic p-forms, then

$$\ker D^+ = \sum_{p \text{ even}} \mathcal{H}^p, \quad \ker D^- = \sum_{p \text{ odd}} \mathcal{H}^p .$$

Because $\dim \mathcal{H}^p = \beta_p$, we obtain

(5) $\text{index } D^+ = \sum (-1)^p \dim \mathcal{H}^p = \chi(M)$,

where $\chi(M)$ is the Euler characteristic.

4. Proving the Index Theorem Using the Heat Equation

There are several different proofs of the index theorem. Those now in favor use the heat equations

$$u_t = - L^*Lu \quad \text{and} \quad u_t = - LL^*u.$$

Let K^1 and K^2 be the heat kernels associated with these two equations (see (1), (6) - (8) in Lecture I). We claim that

(6) $\quad i(L) = \int [K^1(y,y;t) - K^2(y,y;t)] \, dy$

for all $t > 0$, that is,

(7) $\quad i(L) = \text{trace} \left(e^{-tL^*L} - e^{-tLL^*} \right) = \sum_j \left(e^{-\lambda_j t} - e^{-\mu_j t} \right).$

where λ_j are the eigenvalues of L^*L and μ_j those of LL^*. Notice that if $\lambda_j \neq 0$ is an eigenvalue of L^*L, then it is also an eigenvalue of LL^* (since if $L^*L\phi = \lambda\phi$ then $LL^*(L\phi) = \lambda(L\phi)$). Also, the multiplicity of the eigenvalue $\lambda = 0$ of L^*L is dim ker L^*L = dim ker L, with a similar statement for $\mu = 0$. Therefore the non-zero eigenvalue terms in (7) all cancel, while the zero eigenvalue terms give the index of L.

Armed with the formula (6) for the index one needs other properties of the heat kernels K^1 and K^2 to obtain a formula expressing the integrand in (6) in terms of characteristic classes of the manifold. There are several ways of doing this. The method currently under intense study is based on an idea of Witten and investigates the asymptotic behavior of the integrand as $t \to \infty$. But the dust has yet to settle.

5. The Dirac Operator

For the de Rham complex, in (3) above we found an elliptic operator D whose square was the Laplacian. This leads one to seek other operators which are the square root of the Laplacian.

First we work in $\mathbb{R}^n$ with the Laplacian acting on vectors $u = (u_1, \ldots, u_N)$. Thus we seek $N \times N$ constant matrices $E_1, \ldots, E_n$ so that

(8) $\quad \left(\sum_j E_j \frac{\partial}{\partial x_j} \right)^2 = - \Delta I.$

Expanding the left side, we find that

(9) $\quad E_j^2 = - I \quad \text{and} \quad E_i E_j + E_j E_i = 0 \quad \text{for } i \neq j.$

Once one has these matrices, the <u>Dirac operator</u> is defined by

(10) $$D = \sum E_j \frac{\partial}{\partial x_j}$$

and satisfies (8) (actually, Dirac wanted a square root of the wave operator $u_{xx} + u_{yy} + u_{zz} - u_{tt}$, but formally replacing t by iw one is in our case). The Dirac operator is a first order elliptic operator.

Matrices $E_1, \ldots, E_n$ with the multiplication rules (9) generate an algebra, called the <u>Clifford algebra</u>. Given any inner product space V, if $e_1, \ldots, e_n$ are an orthonormal basis, the rules (9) can be summarized as

(11) $u \cdot v + v \cdot u = -2\langle u,v \rangle$

for any vectors $u,v \in V$. Thus, the Clifford algebra C_n over $\mathbb{C}$ can be described abstractly as the tensor algebra generated by $e_1, \ldots, e_n$ divided out by the ideal defined by (11). The $N \times N$ matrices give a representation of C_n as matrices acting on a vector space of dimension N. If n = 2k is even, the algebra is simple, that is, one obtains the whole algebra of $N \times N$ matrices and the N-dimensional space that these matrices act on is called the vector space S of <u>spinors</u>. Abstractly, we have a representation $\rho: C_n \to \text{End}(S)$. Clearly the dimension of C_n is 2^n; so $N^2 = 2^n$; if n = 2k then $N = 2^k$. (In the special case of $\mathbb{R}^n$ with n = 4, then also N = 4 but one should <u>avoid</u> the temptation of identifying the 4-dimensional space S with $\mathbb{R}^4$. A useful exercise is to actually find the matrices $E_1, \ldots, E_4$ in this case.)

This construction has been done for a single inner product space V of dimension n = 2k. It is natural to attempt the same construction on a compact Riemannian manifold (M,g) of dimension 2k, replacing V by the tangent spaces. To carry out the details, one needs to assume M is oriented and that M has a spin structure. The only topological obstruction to the spin structure is the 2nd Stiefel-Whitney class, w_2, so one needs $w_2 = 0$. Using an appropriately adapted connection ∇ on the space of spinors the Dirac operator is written

$$D = \sum_{j=1}^{n} E_j \nabla_j .$$

6. The Lichnerowicz Vanishing Theorem

Since one has a new elliptic operator, the Dirac operator, one should attempt to see if one can again use the ideas in Bochner's vanishing theorem. The corresponding Weitzenböck-type formula is

(12) $\quad D^2 = \nabla^*\nabla + \frac{1}{4}$ scalar curvature ,

where the scalar curvature of (M,g) is the sum of the Ricci curvatures. By identical reasoning as before, we find that if the scalar curvature is positive, then ker D^2 = ker D = 0 (the elements in ker D are called <u>harmonic spinors</u>).

To use this, we need the analogue of the Betti number β_1; this is supplied by the index theorem. Since $D = D^*$ is self-adjoint, then $i(D^2) = 0$. A non-trivial index can be found by a construction motivated by the example of Section 3 above. As our substitute for the spaces Ω^{ev} and Ω^{odd} on a manifold of dimension $2k$, let τ be the Clifford product

$$\tau = i^k E_1 E_2 \ldots E_{2k}$$

(essentially the volume element). By an easy computation $\tau^2 = 1$. Since τ is an element of the Clifford algebra, it acts on the spinors S and has eigenvalues ± 1. Let S^+ and S^- be the corresponding eigenspaces (one can also define $S^{\pm}$ = image of the projector $(1\pm\tau)/2)$). If $\Gamma(S\pm)$ is the space of sections of the spinor bundle, then $D: \Gamma(S^+) \to \Gamma(S^-)$ and $D^-: \Gamma(S^-) \to \Gamma(S^+)$ so we can define D^+ as the restriction of D to $\Gamma(S^+)$, and D^- similarly. Then

$$i(D^+) = \dim \ker D^+ - \dim \ker D^-$$

As a consequence of our observations, if scalar curvature >0 then $i(D^+) = 0$. On the other hand, for an oriented spin manifold of dimension $4k$, the index theorem shows that the index $i(d^+)$ is the $\hat{A}(M)$-genus. Consequently, if $\hat{A}(M) \neq 0$, then M does not admit a metric with positive scalar curvature. This is Lichnerowicz's vanishing theorem. (As a contrast, Aubin has proved that one can always find a metric with negative scalar curvature. Then Kazdan-Warner (see Lecture V below) showed that any function which is negative somewhere is the scalar curvature of some metric.)

Note that if there is no metric with positive scalar curvature, then there is surely no metric with positive Ricci or sectional curvature.

Hitchin later extended Lichnerowicz's argument to show that certain exotic spheres do not have positive scalar curvature metrics. Recently there has been a striking generalization by Gromov-Lawson. One of their results is that if a compact M has a metric with non-positive sectional curvature, then there is no metric with positive scalar curvature. In particular, the torus T^n has no metric with positive scalar curvature. Using quite different techniques (minimal surfaces), Schoen-Yau have also obtained obstructions to the existence of positive scalar curvature metrics.

Lecture IV

1. Nonlinear Elliptic Equations

We shall write a kth order nonlinear differential equation - or a system of such equations - as

(1) $\quad F(x, \partial^k u) = 0$.

The <u>linearization</u> or <u>first variation</u> at the function u is the linear operator

(2) $\quad Lv = \dfrac{d}{dt} F(x, \partial^k(u+tv))\big|_{t=0}$.

The equation (1) is <u>elliptic at</u> (x,u), that is, it is elliptic at x for the function u, if its linearization L at u is elliptic at x.

For example, the linearization of $yu_{xx} + uu_{yy} = 0$ at u is

$Lv = yv_{xx} + uv_{yy} + $ lower order terms .

This is elliptic at the points where both y and $u(x,y)$ have the same sign. The reader should verify that the mean curvature equation

(3) $\quad \nabla \cdot \left(\dfrac{\nabla u}{\sqrt{1+|\nabla u|^2}} \right) = H$

is elliptic for all functions u at all points, while the Monge-Ampère equation

(4) $\quad u_{xx} u_{yy} - u_{xy}^2 = f(x,y)$

is elliptic at a solution $u(x,y)$ precisely at the points where $f(x,y) > 0$.

2. Elliptic Regularity

From analytic function theory, one knows that a function $u \in C^2$ that satisfies $u_{xx} + u_{yy} = 0$ is automatically C^∞, in fact, it is real analytic (i.e. it has a power series expansion in x and y). Solutions of nonlinear elliptic equations (1) enjoy similar regularity properties. Better yet, these properties still hold for <u>overdetermined elliptic systems</u>. These are systems where the symbol need only be injective (an example is the gradient, another example is the Cauchy-Riemann equations in $\mathbb{C}^n$).

One can summarize the results in a table: assume that $u \in C^k$ is an elliptic (possibly overdetermined elliptic) solution of (1) in an open set Ω and that for some integer $j \geq 1$ the function $F(x,s)$ has the following regularity for $x \in \Omega$ and all s (here s represents the variables of the k-jet of u)

$F \in$	C^1	$C^{j+\alpha}$	C^∞	C^ω
then in Ω $u \in$	$C^{k+\lambda}$ all $0<\lambda<1$	$C^{j+k+\alpha}$	C^∞	C^ω

For a solution $u \in H_k^p(\Omega)$ of a <u>linear</u> elliptic equation

(5) $\quad Lu \equiv \displaystyle\sum_{|r|\leq k} a_r(x)\partial^r u = f(x)$,

and $j \geq 0$, assume that the coefficients

and	$a_r \in$	C^j	$C^{j+\alpha}$	C^∞	C^ω
then	$f \in$	H_j^p	$C^{j+\alpha}$	C^∞	C^ω
	$u \in$	H_{k+j}^p	$C^{k+j+\alpha}$	C^∞	C^ω

Typical applications of these are that if $u \in C^2$ satisfies (3) and if $H \in C^\infty$ then $u \in C^\infty$, and also u is real analytic at the points where H is real analytic. Note that regularity is a local property, not a global property.

For many equations, such as $-\Delta u = f(x)$, one can define a <u>weak solution</u> $u \in H_1^2(\Omega)$. As motivation, if $u \in C^2(\Omega)$ is a classical solution, then multiplying $-\Delta u = f$ by $\phi \in C_0^\infty(\Omega)$ and integrating by parts one finds that

$$\int_\Omega \nabla u \cdot \nabla \phi = \int_\Omega f\phi \ .$$

This formula makes sense for any u in $H_1^2(\Omega)$ and any $\phi \in H_1^2(\Omega)$ with $\phi = 0$ on $\partial\Omega$ - in some appropriate sense - to avoid boundary terms in the integration by parts. Thus, if $u \in H_1^2(\Omega)$ satisfies this formula for any such ϕ, we say it is a <u>weak solution</u> of $-\Delta u = f$. In this situation, we need to prove auxiliary regularity results - such as $u \in H_2^2$ - in order to apply the preceding regularity theorem. Note that the idea of a weak solution arises naturally in obtaining the Euler-Lagrange equations of a problem in the calculus of variations, such as for minimal surfaces and harmonic maps.

3. Local Existence for Elliptic Equations

The existence theory for nonlinear elliptic equations consists of some methods and examples. There is no adequate general theory yet. New examples often bring unexpected phenomena. In this and the next lecture we will focus our attention on a few methods and present geometric applications using these methods.

One rather simple-sounding question is if one can find some solution of (1) in a neighborhood of a point $x \in \mathbb{R}^n$. To have some perspective, we point out that the simple linear equation in $\mathbb{R}^2$

(6) $\quad u_x + ixu_y = f(x,y)$

has <u>no</u> solution in any neighborhood of $x = 0$ for most $f \in C^\infty$. This surprising fact was first found by H. Lewy in 1956 (he gave a slightly different example). If f is analytic, one can use power series to find a solution.

However, for elliptic equations (1) with $F(x,s)$ smooth, the story is quite simple: if there is an elliptic solution u_0 at x_0, then there is a solution $u(x)$ in some neighborhood $|x-x_0|<\epsilon$ of x_0. Moreover, one can also specify that $|\partial^j u - \partial^j u_0| < \text{const } \epsilon^{k-|j|+\alpha}$ for all $|j| \leq k$ and $0 < \alpha < 1$.

The proof of this uses the standard implicit function theorem in Banach spaces. By a preliminary change of variables, one may assume that $x_0 = 0$ and $u_0 \equiv 0$, so the solvability at $x_0 = 0$ means $F(0,0) = 0$. For clarity, rewrite (1) as $F(x,\partial^j u) = 0$, $|j| \leq k$. Make the change of scale $x = \lambda y$, $u = \lambda^k v$. Then (1) for $v(y)$ becomes (with $\partial_y = \partial/\partial y$)

$$T(v;\lambda) \equiv F(\lambda y, \lambda^{k-|j|} \partial_y^j v) = 0, \quad |j| \leq k.$$

It is enough to find some $\lambda > 0$ so that we can solve this in the ball $|y| < 1$. Clearly $T(0;0) = 0$. To apply the implicit function theorem we need that the linearization $T_v(0;0)$ is invertible as a map between appropriate Banach spaces. Standard machinery for linear elliptic equations with constant coefficient allows one to complete the proof. The reader may find it instructive to carry out the details for an <u>ordinary</u> differential equation with $T: C^k \times \mathbb{R} \to C^0$ (instead of C^k, it is more convenient to use the subspace of $u \in C^k$ with $\partial^j u(0) = 0$ for $|j| \leq k-1$).

An immediate application is the local solvability of $u_{xx}u_{yy} - u_{xy}^2 = f$ near the origin if $f(0,0) = c^2 > 0$, since $u_0 = c(x^2+y^2)/4$ is an elliptic solution at the origin. It is interesting to point out that if $f(0,0) = 0$, then - even for $f \in C^\infty$ - we do not know if one can always find a solution $u \in C^2$ of this equation in some neighborhood of the origin. A slight modification of this equation arises in the still unresolved question if one can always locally isometrically embed a two-dimensional Riemannian manifold into $\mathbb{R}^3$ (the answer is known to be "yes" at all points where the Gauss curvature is not zero).

4. Complex Structures on $\mathbb{R}^2$

A <u>complex structure</u> is just a way to decide which functions are analytic. One customarily says $f \in C^1$ is analytic if $\partial f/\partial \bar{z} = 0$, that is,

(7) $\quad \left(\dfrac{\partial}{\partial x} + i \dfrac{\partial}{\partial y}\right) f = 0$.

How can we recognize these Cauchy-Riemann equations in other coordinates? In other words, say one is given two real vector fields

$$Q_j = a_j(x,y)\partial/\partial x + b_j(x,y)\partial/\partial y , \qquad j = 1,2,$$

and let $Pf = (Q_1 + iQ_2)f$. Can we find new coordinates $u = u(x,y)$, $v = v(x,y)$ so that in these new coordinates $Pf = 0$ is equivalent to $(\partial/\partial u + i\partial/\partial v)f = 0$?

A necessary condition is clearly that Q_1 and Q_2 be linearly independent. (Equation (6) shows what can happen if Q_1 and Q_2 are dependent.) We claim this is also sufficient. Observe that if we have a solution $w = u + iv$ of $Pw = 0$ with ∇u and ∇v linearly independent, and if we use u and v as new coordinates, then by the chain rule, in these coordinates

$$P = \alpha(u,v)\partial/\partial u + \beta(u,v)\partial/\partial v$$

for some complex-valued functions α and β. But by substitution

$$0 = P(u+iv) = \alpha(u,v) + i\beta(u,v) .$$

Thus $\alpha = -i\beta$ and $P = -i\beta(\partial/\partial u + i\partial/\partial v)$. This proves that $Pf = 0$ if and only if $(\partial/\partial u + i\partial/\partial v)f = 0$. The only gap is that we must locally solve $Pw = 0$. Since Q_1 and Q_2 are linearly independent, one can easily verify that $Pw = 0$ is elliptic so one obtains the local solvability with ∇u and ∇v independent by using the result in the previous section.

5. Complex Structures on $\mathbb{R}^{2n}$

For several complex variables $z_1, \ldots, z_n$, one can similarly ask how one can recognize the Cauchy-Riemann equations

$$\partial f/\partial \bar{z}^1 = \partial f/\partial \bar{z}^2 = \ldots = \partial f/\partial \bar{z}^n = 0$$

in other coordinates. Now we are given complex vector fields

(8) $\quad P_j = \sum\limits_{k=1}^{2n} a_{jk} \, \partial/\partial x^k \quad , \quad j = 1,\ldots,n$

with $P_1,\ldots,P_n, \bar{P}_1,\ldots,\bar{P}_n$ linearly independent and seek a change of coordinates $\zeta = \phi(z)$ so that f satisfies $P_j f = 0$, $j = 1,\ldots,n$ if and only if $\partial f/\partial \bar{\zeta}^k = 0$, $k = 1,\ldots,n$. If we can find these new coordinates, then the P_j will be linear combinations of the $\partial/\partial \bar{\zeta}^k$. Consequently, a necessary condition is that

(9) $\quad [P_j, P_k] = $ linear combination of $P_1, \ldots, P_n$.

Newlander and Nirenberg (1957) proved that the linear independence and the integrability condition (9) are also sufficient that there are coordinates $\zeta^1, \ldots, \zeta^n$ so that $P_j f = 0$ are equivalent to the Cauchy-Riemann equations.

Just as in the simpler case of complex structures on $\mathbb{R}^2$, we will find solutions $\zeta^1,\ldots,\zeta^n$ of $P_j \zeta^k = 0$, $j,k = 1,\ldots,n$, with the gradients of the ζ^j's linearly independent. These will be the new coordinates.

We will sketch Malgrange's proof of this result. He begins with the classical observation that the problem is easy to solve if the coefficients a_{jk} in (8) are real analytic since then one can obtain power series solutions (the integrability conditions (9) are formally just those of the Frobenius theorem). Thinking of $\zeta = (\zeta^1,\ldots,\zeta^n)$ as a complex vector, by a preliminary change of coordinates the equations $P_j \zeta = 0$ may be written as

(10) $\quad P_j \zeta \equiv \dfrac{\partial \zeta}{\partial \bar{z}^j} - \sum\limits_{k} a_{kj} \dfrac{\partial \zeta}{\partial z^k} = 0 \quad , \quad j = 1,\ldots,n$,

where a_{kj} and its first derivatives are zero at the origin. For short we write this as the matrix system

(11) $\quad \dfrac{\partial \zeta}{\partial \bar{z}} = \dfrac{\partial \zeta}{\partial z} A \, , \quad \text{that is,} \quad \zeta_{\bar{z}} = \zeta_z A $.

Because of the special form of (10), the commutators $[P_j, P_k]$ do not involve $\partial/\partial \bar{z}$. Thus the integrability conditions become simply

(12) $\quad [P_j, P_k] = 0$.

The key idea is to introduce new coordinates $w^j = w^j(z,\bar{z})$ in a clever way to be specified shortly, with $w_z(0) = I$, and $w_{\bar{z}}(0) = 0$. In these new coordinates (11) takes the form

(11)' $\quad \zeta_{\bar{w}} = \zeta_w B$

where

(13) $\quad B = (w_z A - w_{\bar{z}})(\bar{w}_{\bar{z}} - \bar{w}_z A)^{-1}$

(note the condition on w at the origin ensures that $\bar{w}_{\bar{z}} - \bar{w}_z A$ is invertible near the origin). Just as in (12), in these new coordinates the integrability conditions for (11)' take the form

(14) $\quad \dfrac{\partial b_{ik}}{\partial w^j} - \dfrac{\partial b_{ij}}{\partial w^k} = \sum_r \left(b_{rj} \dfrac{\partial b_{ik}}{\partial w^r} - b_{rk} \dfrac{\partial b_{ij}}{\partial w^r} \right)$,

where we have written $B = (b_{ij})$.

So far, for any choice of coordinates $w = w(z,\bar{z})$, the system (11)-(12) is entirely equivalent to (11)', (14). Now we pick clever coordinates, requiring that they satisfy the additional conditions

(15) $\quad \sum_k \dfrac{\partial b_{jk}}{\partial w^k} = 0 \quad , \quad j = 1,\ldots,n$.

These equations (14) - (15) for the coefficients b_{ij} as functions of the w^k are an overdetermined elliptic system with analytic coefficients. Therefore the functions b_{ij} are analytic functions of the w and $\bar{w}$. Consequently, the equations (11)',(14) can be solved by using power series to give a solution with $\zeta_w(0) = I$, $\zeta_{\bar{w}}(0) = 0$.

It remains to be show that the functions w^k can be found to satisfy (15). Using (13) and the chain rule (to express $\partial/\partial w$ in terms of $\partial/\partial z$ and $\partial/\partial \bar{z}$), the equations (15) become differential equations for w^k as functions of z and $\bar{z}$. Because A is zero to second order at the origin, this system is elliptic at the origin for the function $w \equiv z$; indeed, the linearized system is

$$-\sum_k \frac{\partial^2 w^j}{\partial z^k \partial \bar{z}^k} + \text{lower order terms} .$$

Since $4\partial^2/\partial z^k \partial \bar{z}^k$ is just the Laplacian, the ellipticity is obvious. The local solvability of (15) for $w = w(z,\bar{z})$ with $w(0) = 0$, $w_z(0) = I$, $w_{\bar{z}}(0) = 0$ is now a consequence of the result earlier in this lecture.

Lecture V

1. Prescribing Curvature

On a given manifold M it is often of interest to find a Riemannian metric g (or possibly just a connection) whose curvature has certain "good" properties. The curvatures studied are usually the sectional curvature, Ricci curvature, or scalar curvature, as well as the mean curvature of a submanifold. One often seeks a metric with constant curvature (recall that a metric with constant Ricci curvature is often called an "Einstein metric") or, as in Yang-Mills theory, one seeks a connection whose curvature minimizes a certain integral.

The existence of these special metrics or connections often helps us understand the geometric or topological properties of a manifold - and the problem of proving the desired existence often leads to interesting and challenging questions in analysis. Of course, one must at the same time keep in mind that there may be topological obstructions to the existence of these structures. Perhaps the simplest such obstruction is provided by the Gauss-Bonnet theorem for a compact 2-manifold:

$$\int_M K dA = 2\pi \chi(M) ,$$

which certainly shows that the sphere, S^2, has no metric with Gausss curvature $K \leq 0$.

In this lecture, all we can hope to do is give some idea of these investigations. We will discuss two questions concerning scalar curvature that lead to elliptic differential equations, and use two standard techniques: the inverse function theorem and the calculus of variations.

2. Negative Scalar Curvature

Let M be a compact manifold of dimension $n \geq 3$. Although there are topological obstructions to the existence of positive or zero scalar curvature metrics (see the end of Lecture III), one can always find a metric g_0 with scalar curvature $S_0 = -1$. Given a function $S(x)$ that is negative somewhere, can one find a metric g whose scalar curvature is S? The answer is "yes", and is due to Kazdan-Warner.

The simplest way to seek this new metric g is in the form $g = p(x)g_0$, where $p(x) > 0$. Then the metrics g and g_0 are <u>pointwise conformal</u>. The formulas become simpler if one writes $p = u^{4/(n-2)}$, where $u(x) > 0$. Then the scalar curvature S of

(1) $\quad g = u^{4/(n-2)} g_0$

is given by the formula

(2) $\quad -\gamma \Delta_0 u + S_0 u = S u^\alpha$, $\qquad u > 0$,

where $\gamma = 4(n-1)/(n-2)$, $\alpha = (n+2)/(n-2)$, and Δ_0 is the Laplacian of g_0 (note our sign convention $\Delta u = +u''$ on $\mathbb{R}$).

For our problem, $S_0 = -1$. If we could always solve the nonlinear partial differential equation (2) with any function S which is negative somewhere, then the metric g of (1) is the desired metric. However, a necessary condition to be able to solve (2) is

(3) $\quad \int_M S \, dx_0 < 0$,

where dx_0 is the volume element for g_0. To obtain this, just multiply (2) by $u^{-\alpha}$ and integrate over M (the easy computation uses one integration by parts).

A deeper investigation shows that even (3) is not sufficient for the solvability of (2). In any case, since (3) is not necessarily satisfied if S is just negative somewhere, one can certainly not solve our original problem by only solving (2). Geometric considerations suggest some additional flexibility: use the group of diffeomorphisms. Thus, instead of (2) we seek a solution of

(4) $\quad -\gamma \Delta_0 u + S_0 u = (S \circ \phi) u^\alpha$, $\qquad u > 0$,

where now the unknowns are both u and a diffeomorphism ϕ. Indeed, if $S \circ \phi = \phi^*(S)$ is the scalar curvature of some metric $g = u^{4/(n-2)} g_0$, then the pulled-back metric $g_1 = (\phi^{-1})^*(g)$ is the metric we want with scalar curvature S. Consequently, we try to solve (4).

Rewrite (2) as

(5) $\quad T(u) \equiv u^{-\alpha}(-\gamma \Delta_0 u + S_0 u) = f$.

Clearly $T(1) = S_0$. By using the inverse function theorem, we thus first try to solve $T(u) = f$ for all $f \approx S_0$, say $\|f - S_0\| < \delta$. The second step will be to seek a diffeomorphism adapted to our given function S so that $\|S \circ \phi - S_0\| < \delta$, since then we can solve (4) and complete the problem.

Now to use the inverse function theorem, we need only check that the linearization $L = T'(1)$ is invertible. By an easy computation

$$Lv = \frac{d}{dt}T(1+tv)\Big|_{t=0} = -\gamma\Delta_0 v - (\alpha-1)S_0 v \ .$$

Since this is self-adjoint and elliptic, the Fredholm alternative, i.e. the basic decomposition (7) or (7)', shows L is invertible if its kernel is zero. But because $\alpha > 1$ and $S_0 < 0$, so $(\alpha-1)S_0 < 0$, this is obvious (see Example 1 in Lecture II.2). Therefore by the inverse function theorem T maps a neighborhood of $u = 1$ onto a neighborhood $\|f-S_0\| < \delta$ of S_0. By making u near 1 we guarantee that $u > 0$. Regularity for (5) assures us that if $f \in C^\infty$ then $u \in C^\infty$.

We must choose function spaces. The standard candidates are

$$T: C^{2+\alpha} \to C^\alpha \quad \text{and} \quad T: H_2^p \to L^p \ .$$

If we use the Sobolev space H_2^p we must be a bit careful because of the non-linear term $u^{-\alpha}$. From (5) we would like $u^{-\alpha}$ to be a continuous function. By the Sobolev theorem (end of Lecture I), if $p > n$ and $u \in H_1^p$ then u is continuous, so surely if $p > n$ and $u \in H_2^p$ then u is continuous. In fact, if $p > n/2$ and $u \in H_2^p$ then u is continuous. Thus, if we use $T: H_2^p \to L^p$ we require that $p > n/2$.

The final choice of our space hinges on the second step, where we seek a diffeomorphism ϕ so that $\|S_0\circ\phi - S_0\| < \delta$, i.e. $\|S_0\circ\phi + 1\| < \delta$. This is hopeless if we use the uniform norm - or even worse the C^α norm. Thus we try Sobolev spaces and want ϕ so that

(6) $\quad \|S_0\circ\phi - (-1)\|_{L^p} < \delta$.

Assume first that S equals -1 somewhere, and hence S is very near -1 on an open set. Now we simply find a diffeomorphism that spreads this open set over so much of M that (6) is satisfied. We thus use H_2^p for any $p > n/2$.

In the general case where S is only known to be negative somewhere, we multiply S by a constant to have $S_1 = cS$ take on the value -1. Then use the above construction to find a metric with scalar curvature S_1. Scaling this metric gives a metric with scalar curvature S.

For this construction we began with a metric g_0 having $S_0 = -1$. One can attempt to carry out the same procedure if $S_0 = 0$ (as on the torus) or S_0, a positive constant (as on the sphere). However, the linearization is not necessarily invertible. One gets around this by slightly perturbing the initial metric. Although there are some technical difficulties, the end result gives a similar conclusion.

It is useful to note that although all the geometric data in this problem were smooth, we were forced to use the L^p norm because of the additional flexibility they give to allow us to find a diffeomorphism so that (6) is satisfied.

3. The Yamabe Problem

For a two-dimensional Riemannian manifold the uniformization theorem gives a conformal metric with constant Gaussian curvature. Perhaps the simplest analogue in higher dimensions is to seek a conformal metric with constant scalar curvature. Thus, given (M, g_0) we seek a solution of (2) with $S = \lambda = $ constant, that is

(7) $\quad Lu \equiv -\gamma \Delta_0 u + S_0 u = \lambda u^\alpha$, $\quad\quad u > 0$.

Yamabe asked this question, but his paper had a serious error.

By a preliminary reduction, we first show it is sufficient to consider the three cases $S_0 > 0$, $S_0 \equiv 0$, $S_0 < 0$. Let λ_1 be the lowest eigenvalue of L. By general theory, the corresponding eigenfunction, v, is never zero, so we may assume $v > 0$. Then

$$Lv = \lambda_1 v = (\lambda_1 v^{1-\alpha}) v^\alpha .$$

From (1)-(2) the conformal metric $g_1 = v^{4/(n-2)} g_0$ has scalar curvature $S_1 = \lambda_1 v^{1-\alpha}$, which has the same sign as λ_1. This completes the preliminary reduction and we can assume $S_0 > 0$, $S_0 \equiv 0$, or $S_0 < 0$.

The simplest case is $S_0 \equiv 0$, since we can just take $u \equiv 1$. Next simplest is $S_0 < 0$. If $c > 0$ is a constant, then $Lc = S_0 c$. Consequently, there are constants $0 < c_- < c_+$ so that $Lc_- \leq -(c_-)^\alpha$ and $Lc_+ \geq -(c_+)^\alpha$ (recall $\alpha > 1$). Thus, the constant functions $u_\pm = c_\pm$ are __super__ and __subsolutions__ of (7) with $\lambda = -1$. Because $0 < u_- < u_+$, a general construction based on the maximum principle guarantees the existence of a solution $0 < u_- \leq u \leq u_+$ of (7) (still with $\lambda = -1$) and completes the case $S_0 < 0$. One can easily prove this solution is unique by the maximum principle.

If $S_0 > 0$, one must work much harder. Motivated by the resemblance of (7) to finding eigenvalues of L (i.e. the case $\alpha = 1$), one attempts to find a solution to (7) by finding a critical point of the Yamabe functional

(8) $\quad Y(u) = \int \left[\gamma |\nabla_0 u|^2 + S_0 u^2 \right] dx_0$,

where u satisfies the constraint

(9) $\quad \int u^{\alpha+1} \, dx_0 = 1$.

You should check that (7) is the Euler-Lagrange equation for (8)-(9).

Because $S_0 > 0$, the functional $Y(u)$ is essentially the norm in $H_1^2(M)$. In fact, it is obviously equivalent to the norm so we may simply use $\|u\|^2 = Y(u)$ as the H_1^2 norm. Clearly $Y(u) > 0$ so it is natural to seek a minimum of Y. Let

$$\sigma = \inf Y(u) = \inf \|u\|_{H_1^2}$$

for all $u \in H_1^2$ that satisfy (9). Then there are u_j so that $Y(u_j) \to \sigma$.

But we must be careful to check that (9) is defined for all $u \in H_1^2$. By the Sobolev inequality, $u \in L^p$ for all $p \leq 2n/(n-2)$. Since $\alpha = (n+2)/(n-2)$, then $\alpha + 1 = 2n/(n-2)$ so (9) does make sense - just barely (if α were any larger, this would have failed). Of course, as we remarked at the end of Lecture I, one uses a conformal map $x \to \lambda x$ to find the critical Sobolev exponent so one expects that exactly this exponent will arise in considering conformal metrics.

Recall that in a Hilbert space, such as H_1^2, a sequence of vectors x_j <u>converges weakly</u> to x if $\langle x_j - x, z \rangle \to 0$ for any vector z (the simplest example is where x_j are an orthonormal basis, in which case $x_j \to 0$ weakly). The virtue of weak convergence is that closed and bounded sets are weakly compact. Moreover, the norm is lower semi-continuous under weak convergence (proof: if $x_j \to x$ weakly, then

$$\|x\|^2 = \lim \langle x, x_j \rangle \leq \lim \inf \|x\| \, \|x_j\|) \ .$$

Applied to our case, since $Y(u_j) \to \sigma$, then $\|u_j\|_{H_1^2} = Y(u_j) \leq$ const so by weak compactness of this bounded set, a subsequence - which we relabel u_j - converges weakly to some u in H_1^2. Because $Y(u_j) = Y(|u_j|)$, we may assume $u_j > 0$. The lower semi-continuity gives

(10) $\quad Y(u) = \|u\|_{H_1^2}^2 \leq \lim \inf Y(u_j) = \sigma.$

<u>If</u> we can show that u also satisfies the constraint (9), then we will have that $Y(u) \geq \sigma$. Combined with (10) this yields $Y(u) = \sigma$ so $u \geq 0$ will be our desired minimum of (8), (9). Since $u \in H_1^2$ we will then have to show that u actually is a smooth function; then the proof will be completed by

improving $u \geq 0$ to the stronger inequality $u > 0$ we need. To prove that $u > 0$, one uses the maximum principle for (7) to show that either $u > 0$ or else $u \equiv 0$. This last possibility is excluded by the constraint (9). The smoothness of u is a little technical since one is at the limit of the Sobolev embedding.

The main difficulty is showing that u satisfies the constraint (9), since so far we only know that $u_j \to u$ weakly in H_1^2. Now by the Sobolov embedding theorem, $H_1^2 \hookrightarrow L^p$ for all $p \leq \alpha + 1$ and this embedding is compact if $p < \alpha + 1$. Thus, the $u_j \to u$ strongly (i.e. in norm) in L^p for all $p < \alpha + 1$ but we are at the limiting case so a much more delicate analysis is needed. It is conceivable that $u \equiv 0$ - and this actually does occur in some closely related problems. One thinks of u_j as concentrating at one point, like a bubble, and the bubble "pops" as $j \to \infty$.

To show that (9) is satisfied, one shows that there is a certain critical constant σ_0 (essentially the norm of the embedding $H_1^2 \hookrightarrow L^{1+\alpha}$) so that one always has $\sigma \leq \sigma_0$ (here σ_0 depends only on the dimension, but not on the metric g or on M). More important, if $\sigma < \sigma_0$ then Aubin (in 1975) proved that the minimizing sequence u_j actually converges strongly to u in H_1^2, and hence strongly in L^p for all $p \leq \alpha + 1$. In particular, the limit function u satisfies the constraint (9). Aubin also showed that in most cases when $n \geq 6$ one actually has $\sigma < \sigma_0$.

However, by using Möbius transformations on S^n, there are situations where $\sigma = \sigma_0$, so until last year the general case was unclear.

Recently, R. Schoen has completed the Yamabe problem by showing that indeed $\sigma < \sigma_0$ except in the trivial case of the sphere S^n with a metric conformal to the standard metric (where one can use the standard metric to get constant curvature). For Schoen's proof, one need only find some function v satisfying the constraint (9) with $Y(v) < \sigma_0$, since then $\sigma = \inf Y < \sigma_0$. But one must choose a very clever function. Schoen essentially uses Green's function for L, and his proof involves some deep information on its behavior near its singularity.

To summarize this section, we find that Yamabe's problem can always be solved: there is a conformal metric with constant scalar curvature.

The significant of these results is that a number of problems in geometry (minimal surfaces, harmonic maps) and physics (Yang-Mills fields)

also lead to problems where there is a loss of compactness because one
is at the critical exponent in the Sobolev inequalities. For some of these
problems we have non-existence results, for others, existence theorems.
Gradually, the picture is being clarified, and seems to be closely related
to the basic geometry of the underlying manifold. The worst cases are manifolds like the sphere where there is a non-compact group operating (as the
conformal transformations - essentially $x \to \lambda x$). Recent work by Bahri-Coron
should also be significant in our understanding.

REFERENCES

There are a few general references.

[1] AUBIN, T., *Nonlinear analysis on manifolds. Monge-Ampère equations*. Die Grundlehren der Math., Vol 252, Springer-Verlag, New York, 1982.

[2] GILBARG, D. and TRUDINGER, N., *Elliptic Partial Differential Equations of Second Order*, 2nd edition, Die Grundlehren der Math., Vol. 224, Springer-Verlag, New York, 1983.

[3] KAZDAN, JERRY L., *Prescribing the Curvature of a Riemannian Manifold*, CBMS Regional Conference Series in Math., No. 57, American Math. Soc., Providence, R.I., 1985.

Lecture I. See [1] and [2] for the basic facts on Hölder and Sobolev spaces. Another reference is the recent text

[4] BRÉZIS, H., *Analyse Fonctionnelle*, Mason, Paris, 1983.

Lecture II. In addition to [1], [2], [4], a completely self-contained proof of the L^2 theory for the Hodge Laplacian on compact manifolds is presented in

[5] WARNER, F.W., *Foundations of Differentiable Manifolds and Lie Groups*, Springer-Verlag, New York, 1984 (reprinted from the 1971 edition published by Scott-Foresman).

The L^p and Hölder space estimates for elliptic systems can be found in

[6] MORREY, C.B., *Multiple Integrals in the Calculus of Variations*, Die Grundlehren der Math., Vol. 130, Springer-Verlag, New York, 1966.

Lecture III. A useful reference for the Atiyah-Singer index theorem is the expository lecture

[7]   ATIYAH, M.F., "The Heat Equation in Riemannian Geometry", Seminaire Bourbaki 1973/1974, Exp. 436, Lecture Notes in Math., Vol.431, Springer-Verlag, Berlin, 1975.

The Bochner and Lichnerowicz vanishing theorems, and generalizations, are discussed in the survey article

[8]   WU, H., "The Bochner Technique", Proc. 1980 Bejing Sympos. on Diff. Geom. and Diff. Eq., Vol. 2, (S.S. Chern and Wu Wen-tsun, editors), Science Press, China, and Gordon & Breach, New York, 1982, 929-1072,

and in

[9]   GROMOV, M., and LAWSON, H.B., "Positive curvature and the Dirac opeator on complete Riemannian manifolds", Inst. Hautes Études Sci. Publ. Math., 59 (1983), 83-196.

Lecture IV. The facts on nonlinear elliptic equations are in [2] and [6]. Our discussion of Complex Structures closely follows

[10] NIRENBERG, L., Lectures on Linear Partial Differential Equations, C.B.M.S. Regional Conference Series in Math., No. 17, Amer. Math. Soc., Providence, R.I., 1973.

Lecture V. See [3] for a more detailed survey and bibliography on these topics.

METRIC DIFFERENTIAL GEOMETRI

by

Karsten Grove

This is essentially an exposition of a series of five lectures given at the 1985 Nordic Summer School in Lyngby, Denmark.

The main purpose of the lectures is to give an impression of the part of riemannian geometry that relates most directly with metric properties. The first two lectures are supposed to serve also as a general preparation. Given the limits we have tried to make the lectures as self-contained as possible. When a non-obvious proof is omitted a reference is given, or it can be found in one of the basic references [KN], [BC], [GKM], [CE], or [K1]. We have tended to include proofs when it is helpful for the general understanding and/or when the same proof cannot be found explicitly in the literature.

Here is an outline. The first lecture gives a brief treatment of bundles and general connections. In Particular, the equivalence of various difinitions of connections is discussed. We use [KN] or [BC] as general references. In the second lecture we review the basics of riemannian geometry. The third lecture provides the foundations of metric differential geometry. We give short proofs of the Rauch comparison theorems, the Toponogov comparison theorem, and the Bishop-Gromov comparison theorem. The fourth lecture is devoted to the general area of relations between geometry and topology of riemannian manifolds. Our treatment here is naturally far from complete. We have chosen to limit our discussion to manifolds with lower curvature bounds only, in particular to manifolds with non-negative curvature. For recent spectacular progress on manifolds with non-positive curvature we refer to work of Ballmann, Brin, Burns, Eberlein and Spatzier (cf. [BBE], [BBS], [BA], [BS_1] and [BS_2]. Here dynamics plays an essential role beside comparison theory. Another is the "almost flat" theorem of Gromov [G] (cf. also [BK]). This is an ultimate application of comparison theory. In the last lecture we present some fundamental ideas of Gromov relating to metric space structures on the set of all riemannian manifolds. We illustrate this viewpoint by some applications to finiteness theorems and pinching theorems.

1. BUNDLES AND CONNECTIONS.

This section is devoted to a brief discussion of the fundamental notions of bundles and connections in bundles.

In general, a <u>bundle</u> with <u>fiber</u> F, <u>total space</u> E and <u>base space</u> M is a map $\pi : E \to M$ where each point $p \in M$ has an open neighborhood $U \subset M$ such that $\pi : \pi^{-1}(U) \to U$ up to diffeomorphism is the projection $pr_1 : U \times F \to U$; i.e. there is a diffeomorphism $\Phi : \pi^{-1}(U) \to U \times F$ which makes the diagram

(1.1)
$$\pi^{-1}(U) \xrightarrow{\Phi} U \times F$$
$$\pi \searrow \swarrow pr_1$$
$$U$$

commutative. Such a diffeomorphism is called a <u>trivialization</u> of the bundle $E \to M$ over U.

Note that local trivializations Φ_α, Φ_β with $U_\alpha \cap U_\beta \neq \emptyset$ determine <u>transition functions</u>.

(1.2) $$\varphi_{\beta\alpha} : U_\alpha \cap U_\beta \to \text{Diff}(F)$$

(1.3) $$\Phi_\beta \circ \Phi_\alpha^{-1}(p,f) = (p, \varphi_{\beta\alpha}(p)(f)), \quad (p,f) \in \left(U_\alpha \cap U_\beta\right) \times F.$$

Clearly

(1.4) $$\varphi_{\alpha\alpha}(p) = \text{id}_F, \quad p \in U_\alpha,$$

and

(1.5) $$\varphi_{\gamma\beta}(p) \circ \varphi_{\beta\alpha}(p) = \varphi_{\gamma\alpha}(p), \quad p \in U_\alpha \cap U_\beta \cap U_\gamma.$$

It is useful to observe that $\pi : E \to M$ may be reconstructed from the projections $pr_1 : U_\alpha \times F \to U_\alpha$, $\bigcup_\alpha U_\alpha = M$ by means of transition functions. To get E from the disjoint union $\bigsqcup_\alpha U_\alpha \times F$ simply identify $(p_\alpha, f_\alpha) \in U_\alpha \times F$ with $(p_\beta, f_\beta) \in U_\beta \times F$ if and only if $p_\alpha = p_\beta = p \in U_\alpha \cap U_\beta$ and $f_\beta = \varphi_{\beta\alpha}(p)(f_\alpha)$.

Normally in geometry, the fiber F will carry a certain structure which is preserved by transition functions for suitable choices of local trivializations. As a matter of fact, much of geometry and topology is related to problems on existence and properties of such "geometric" structures (cf. [Cn]). It is a beautiful fact of nature that very often the group of "structure preserving" transformations form a finite dimensional Lie group. As a consequence much of the study of bundles is confined to bundles which admit local trivializations whose transition functions take values in a (fixed) finite dimensional Lie group $G \subset \text{Diff}(F)$ considered as a group of transformations acting from the <u>left</u> on F. This group is called the <u>structure group</u> of $E \to M$.

A <u>vector bundle</u> is a bundle $E \to M$ with fiber $F = V$ a vector space and structure geoup $G \subset G\ell(V)$ a subgroup of the group of linear isomorphisms of V.

A <u>principal bundle</u> is a bundle $P \to M$ with fiber $F = G$ a Lie group and structure group a subgroup of G considered as a group of left translations on G. Right multiplication on G induces via local trivialization a <u>right action</u>.

(1.6) $\qquad\qquad P \times G \to P \; ; \; (p \cdot g) \cdot h = p \cdot (g \cdot h)$

which is clearly <u>free</u>, i.e. $p \cdot g = p$ iff $g = e$. Moreover the bundle projection $P \to M$ is the quotient map $P \to P/G = M$ onto the orbit space of this action.

1.7 Examples.

(i) The projection $S^n \to P^n(R)$ of the n-sphere to the real projective space is a principal bundle with group $G = O(1) = Z_2$.

(ii) The Hopf map $S^{2n+1} \to P^n(C)$ to the complex projective space is a principal bundle with group $G = U(1) = S^1$.

(iii) The Hopf map $S^{4n+3} \to P^n(H)$ to the quaternionic projective space is a principal bundle with group $G = Sp(1) = S^3$.

(iv) If $E \to M$ is a vector bundle with fiber V the bundle of bases of E, $B(E) \to M$ is a principal bundle with group $G\ell(V)$.

The last example is a special case of a general construction of a principal G-bundle $P \to M$ from a bundle $E \to M$ with structure group G. P is simply obtained from the disjoint union $\bigsqcup_\alpha U_\alpha \times G$ by identifying $(p_\alpha, g_\alpha) \in U_\alpha \times G$ with $(p_\beta, g_\beta) \in U_\beta \times G$ iff $p_\alpha = p_\beta = p \in U_\alpha \cap U_\beta$ and $g_\beta = \varphi_{\beta\alpha}(p) \cdot g_\alpha$.

Conversely given a principal G-bundle $P \to M$ and a manifold F on which G acts on the left

$$G \times F \to F \quad ; \quad h \cdot (g \cdot f) = (h \cdot g) \cdot f \quad ; \quad e \cdot f = f$$

one constructs an <u>associated bundle</u> $E \to M$ with fiber F as follows:

The map

$$P \times F \times G \to P \times F \quad , \quad ((p,f),g) \to (p \cdot g, g^{-1} \cdot f)$$

defines a free right action on $P \times F$. The obvious projection $P \times F \to P \to M$ is invariant under this G-action and the induced projection

$$E = P \times_G F = P \times F/_G \to M$$

is a bundle with fibre $G \times_G F = G \times F/_G = F$ and structure group G. If $\varphi_{\beta\alpha} : U_\alpha \cap U_\beta \to G$ are transition functions for $P \to M$ they are also transition functions for $E \to M$ via the left action of G on F.

It is easy to see that passing from the bundle $E \to M$ with fiber F and structure group G to its principal G-bundle $P \to M$ and then to the associated bundle with fiber F as described above gives back the bundle $E \to M$.

Examples 1.8.

(i) The canonical line bundles (real, complex and quaternionic) over the projective spaces $P^n(R)$, $P^n(C)$ and $P^n(H)$ are the associated bundles of the principal bundles in 1.7 (i) - (iii) via the canonical actions of $O(1), U(1)$ and $Sp(1)$ on R, C and H resp.

(ii) Let $E \to M$ be a bundle with fiber F and structure group G and consider a map $f : N \to M$. The <u>pull-back</u> of $E \to M$ by f is a bundle $f^*E \to N$ with fiber F and structure group G obtained as follows

(1.9)

$$\begin{array}{ccc}
& N \times E & \\
& \overset{\tilde{F}}{\searrow} & \\
f^*E & \longrightarrow & E \\
\text{pr}_1 \downarrow & & \downarrow \pi \\
N & \longrightarrow & M \\
& f &
\end{array}$$

where $f^*E = \{(q,u) \in N \times E \mid f(q) = \pi(u)\}$.

If $\varphi_{\beta\alpha} : U_\alpha \cap U_\beta \to G$ are transition functions for $E \to M$ then $\psi_{\beta\alpha} = \varphi_{\beta\alpha} \circ f$ $f^{-1}(U_\alpha) \cap f^{-1}(U_\beta) \to G$ are transition functions for $f^*E \to N$.

We will now turn to connections in principal G-bundles.

Let $\pi : P \to M$ be a principal G-bundle. The subbundle (distribution) $V \to P$ of the tangent bundle $TP \to P$ of P defined by

(1.10) $$V = \{X \in TP \mid \pi_* X = 0\}$$

is called the <u>vertical</u> bundle of P. Note that $V_u \subset T_u P$, $u \in P$ is the tangent space at u of the fiber, $P_{\pi(u)}$, of $\pi : P \to M$ over $\pi(u) \in M$. There is, however, no canonical complement H_u to V_u i.e. a subspace $H_u \subset T_u P$ with

$$V_u \oplus H_u = T_u P .$$

Such a space is called a <u>horizontal</u> space at u.

A <u>connection</u> in P is a <u>G-invariant horizontal subbundle</u> (distribution) $H \to P$ <u>of</u> $TP \to P$ i.e.

$$V_u \oplus H_u = T_u P , \quad u \in P$$

(1.11)

$$H_{u \cdot g} = (R_g)_* H_u , \quad u \in P , g \in G .$$

Here $(R_g)_*$ is the differential of the right action $R_g(u) = u \cdot g$ of $g \in G$ on P (cf. 1.6).

Vectors $T \in V$ are called vertical and vectors $X \in H$ are called horizontal. Clearly any vector field Z on P can be written uniquely as $Z^v + Z^h$ in vertical and horizontal components. Moreover for each $X_p \in T_p M$ and $u \in P$ there is a unique $\tilde{X}_u \in H_u$ called the <u>horizontal lift</u> of X_p to u.

The choice of a horizontal subspace H_u is of course equivalent to the choice of a projection $Q_u : T_u P \to V_u$ (with kernel H_u). This can be pushed further since each V_u may be canonically identified with the Lie algebra g of G via the action of G on P. The fiber containing $u \in P$ coincides with the G-orbit through u, which in turn is the image of the imbedding

$$G \longrightarrow P \, ; \quad g \longrightarrow ug .$$

The differential of this map is then a linear isomorphism between $T_e G$ and V_u. In fact these maps define a trivialization of the vertical bundle $V \to P$. The vertical vector fields $\bar{T}$ on P obtained this way from left invariant vector fields $T \in g \simeq T_e G$ are called <u>action fields</u>.

This gives a computationally very useful interpretation of connections:
A connection in $P \to M$ is a <u>1-form</u> ω <u>on</u> P <u>with values</u> in g such that

$$\omega(\bar{T}) = T \quad , \quad T \in g$$

$$R_g^*(\omega) = \mathrm{Ad}(g^{-1}) \circ \omega .$$

The first property expresses that ω is "the identy" on vertical vectors and the second that ω is equivariant. Here $\mathrm{Ad} : G \to G\ell(g)$ is the adjoint representation of G i.e. $\mathrm{Ad}(g^{-1}) : g \to g$ is the differential of $\mathrm{ad}(g^{-1}) : G \to G$, $h \to g^{-1}hg$ taken at $e \in G$.

Note that if $f^*P \to N$ is the pull back of a principal G-bundle $P \to M$ by $f : N \to M$ (c.f. (1.8 iii)) and ω is a connection form for $P \to M$ then $\tilde{F}^*(\omega)$ is a connection form for $f^*P \to N$.

Let in particular $N = [a,b] \subset \mathbb{R}$ and $\gamma : [a,b] \to M$ be a smooth curve. A connection in $P \to M$ then induces a connection in the pull back bundle $\gamma^*P \to [a,b]$. If $\tilde{D}$ is the horizontal lift of the vector field $D = \frac{\partial}{\partial t}$ on $[a,b]$, then each maximal integral curve of $\tilde{D}$ has domain $[a,b]$. The image of these integral curves by Γ in P (c.f. (1.9)) are <u>horizontal lifts</u> of γ i.e. the velocity vectors of these curves are all horizontal It follows that for each $u \in P_{\gamma(a)}$ there is an unique horizontal lift $\tilde{\gamma}$ with $\tilde{\gamma}(a) = u$ and the map defined this way from $P_{\gamma(a)}$ to $P_{\gamma(b)}$ is a diffemorphism of fibers. This diffeomorphism is called <u>parallel transport</u> along γ from $\gamma(a)$ to $\gamma(b)$.

The path dependence of parallel transports is measured by curvature.

The <u>curvature form</u> Ω of a connection form ω is the g-valued 2-form on P defined by

(1.12) $$\Omega(X,Y) = d\omega(X^h, Y^h)$$

where d is the exterior differential. The <u>structural equation</u> states that

(1.13) $$d\omega(X,Y) = -\tfrac{1}{2}[\omega(X), \omega(Y)] + \Omega(X,Y) ,$$

where $[\,,\,]$ denotes the Lie bracket in g. The proof is standard and will not be given here.

Consider now an associated bundle $E = P \times_G F \to M$ with fiber F and principal G-bundle $P \to M$. A connection in $P \to M$. A connection in $P \to M$ defines via the quotient map $P \times F \to P \times_G F$ a <u>horizontal</u> subbundle $H \to E$ of $TE \to E$ in the obvious sense. Likewise horizontal lifts of curves in M to P define via the quotient map $P \times F \to E$ horizontal lifts to E. This in turn defines the notion of parallel transport in $E \to M$ along curves in M.

Now suppose $E \to M$ is a d-dimensional real vector nundle and $B(E) \to M$ its principal $G\ell(d,R)$-bundle of bases of E. Given a connection in $B(E) \to M$ and a curve $\gamma : [0,1] \to M$, we can characterize the horizontal lift of γ through $\eta(0) \in E_{\gamma(0)}$ as a section $\eta : [0,1] \to E$ along γ as follows: If $b \in B(E)_{\gamma(0)}$ and $\tilde{\gamma}$ is the horizontal lift of γ to $B(E)$ through b, then η has constant coefficients w.r.t $\tilde{\gamma}$. We will say that η is <u>parallel</u> along γ and that $\tilde{\gamma}$ is a <u>parallel basis</u> along γ.

We conclude this section by showing that a connection in a vector bundle $\pi : E \to M$ is equivalent to a certain differential operator ∇.

In order to define this operator we identify the vertical space $V_u = T_u(E_{\pi(u)}) \subset T_u E$ with the fiber $E_{\pi(u)}$ over $\pi(u)$ in the usual way. Given a section $\eta \in C^\infty(E)$ in $\pi : E \to M$ i.e. $\pi \circ \eta = id_M$ we define a linear map

$$\nabla \eta : T_p M \to E_p$$

for each $p \in M$ by

(1.14) $\qquad \nabla \eta (X) = \nabla_X \eta = \eta_*(X)^V = \eta_*(X) - \eta_*(X)^h$

for $X \in T_p M$. If X is represented by a curve γ on M i.e. $\dot{\gamma}(0) = X$ then $\eta_*(X)$ is represented by the section $\eta \circ \gamma$ along γ. Moreover $\eta_*(X)^h$ is represented by the horizontal lift of γ to E through $\eta(p)$ i.e. by $\sum_{i=1}^{d} c_i e_i(t)$, where $\tilde{\gamma}(t) = \{e_i(t)\}$ is a horizontal lift (parallel basis) of γ to $B(E)$ and $\eta(p) = \sum_{i=1}^{d} c_i e_i(0)$. Since $\eta(\gamma(t)) = \sum_{i=1}^{d} f_i(t) e_i(t)$ with $f_i(0) = c_i$ we get from (1,14) that

(1.15) $\qquad \nabla_X \eta = \sum_{i=1}^{d} f_i'(0) e_i(0)$.

The operator

$$\nabla : C^\infty(E) \to C^\infty(L(TM;E)) = C^\infty(TM^* \oplus E)$$

defined above is called a <u>covariant derivative</u> for the bundle $E \to M$. It follows directly from (1.14) and (1.15) that

$$\nabla : C^\infty(TM) \times C^\infty(E) \to C^\infty(E)$$

satisfies the identities

(1.16)   (i) $\nabla_{X_1+X_2} \eta = \nabla_{X_1} \eta + \nabla_{X_2} \eta$

(ii) $\nabla_{fX} \eta = f \nabla_X \eta$

(iii) $\nabla_X (\eta_1 + \eta_2) = \nabla_X \eta_1 + \nabla_X \eta_2$

(iv) $\nabla_X (f \eta) = X(f) \cdot \eta + f \nabla_X \eta$.

Conversely given ∇ satisfying (1.16) one obtains a connection in $B(E) \to M$ whose covariant derivative is ∇.

To see this first observe that

$$\nabla_{X_1} \eta(p) = \nabla_{X_2} \eta(p) \quad \text{if} \quad X_1(p) = X_2(p)$$

and

$$\nabla_X \eta_1|_U = \nabla_X \eta_2|_U \quad \text{if} \quad \eta_1|_U = \eta_2|_U \ .$$

In local trivializations Ψ and Φ of $TM \to M$ and $E \to M$ over $U \subset M$ determined by coordinate vector fields $\frac{\partial}{\partial x_1}, \ldots, \frac{\partial}{\partial x_n}$ and linearly independent sections $\eta_1, \ldots, \eta_k$ resp. we may write

(1.17) $\quad \nabla_{\frac{\partial}{\partial x_i}} \eta_j = \sum_{k=1}^{d} \Gamma_{ij}^k \eta_k \ , \quad i = 1, \ldots, n \ ; \ j,k = 1, \ldots, d \ .$

The functions $\Gamma_{ij}^k : U \to \mathbb{R}$ are called the components of ∇ w.r.t. the trivializations Ψ and Φ.

If $\eta : [0,1] \to E$ is a section along $\gamma : [0,1] \to U$ one can define the <u>covariant derivative</u> $\nabla_D \eta$ of η along γ. In local coordinates one has

(1.18) $\quad \nabla_D \eta (t) = \sum_k \left(a_k'(t) + \sum_{ij} \Gamma_{ij}^k (\gamma(t)) \gamma_i'(t) a_j(t) \right) \eta_k (\gamma(t))$

where $\eta(t) = \sum_k a_k(t) \eta_k (\gamma(t))$ and $(\gamma_1, \ldots, \gamma_n)$ are the coordinate functions of γ in the coordinate system $(x_1, \ldots, x_n)$ on U. The equation $\nabla_D \eta = 0$ is a 1st order linear differential equation. Hence for each initial value $\eta_0 \in E_{\gamma(0)}$ there is a unique η with $\eta(0) = \eta_0$ and $\nabla_D \eta = 0$. This defines the notation of parallel sections along curves by means of ∇. If now $b \in B(E)_{\gamma(0)}$ is a basis for $E_{\gamma(0)}$ there is an unique parallel basis $\tilde{\gamma}$ along γ with $\tilde{\gamma}(0) = b$. The horizontal space $H_b \subset T_b B(E)$ is now defined as the subspace of vectors at b represented by curves of the form $\tilde{\gamma}$ just described.

The <u>curvature tensor</u> of ∇ is defined by

(1.19) $\qquad R(X,Y)\eta = \nabla_X \nabla_Y \eta - \nabla_Y \nabla_X \eta - \nabla_{[X,Y]} \eta$

for $X, Y \in C^\infty(TM)$ and $\eta \in C^\infty(E)$. For each $p \in M$ this defines a trilinear map

$$R_p : T_p M \times T_p M \times E_p \to E_p .$$

In terms of the curvature 2-form Ω (1.12) this is given by

$$R_p(X,Y)\eta = b\Omega(\bar{X}, \bar{Y}) b^{-1}(\eta) ,$$

where $b \in B(E)_p$ is considered as a linear isomorphism $b : R^d \to E_p$ and $\bar{X}, \bar{Y} \in T_b B(E)$ are horizontal lifts of $X, Y \in T_p M$. Note that the Lie algebra of $G\ell(d,R)$ is canonically isomorphic to the space of all dxd-matrices. We will neither use nor prove this identity here.

A connection on a manifold M is by definition a connection, ∇ on the tangent bundle $TM \to M$. In this case a curve $\gamma : [a,b] \to M$ is called a <u>geodesic</u> iff its velocity field $\dot{\gamma} : [a,b] \to TM$ is parallel along γ i.e. $\nabla_D \dot{\gamma} = 0$. In local coordinates (cf. (1.17), (1.18)) one has

$$\nabla_D \dot{\gamma} = \sum_{k=1}^n \left(\gamma_k'' + \sum_{ij=1}^n \Gamma_{ij}^k \circ \gamma \cdot \gamma_i' \cdot \gamma_j' \right) \frac{\partial}{\partial x_k} \circ \gamma = 0$$

which is a quadratic second order equation in $(\gamma_1, \ldots, \gamma_n)$. It follows that for each $X \in TM$ there is a unique maximal geodesic $\gamma_X : J_X \to M$ with $0 \in J_X \subset R$ and $\dot{\gamma}_X(0) = X$. Moreover $\mathcal{O} = \{X \in TM \mid 1 \in J_X\}$ is an open "star shaped" neighborhood of the zero section in $TM \to M$ and the <u>exponential map</u>

(1.21) $\qquad \exp : \mathcal{O} \to M , \quad X \to \gamma_X(1)$

is smooth.

Note that the line $t \to tX \in T_p M \cap \mathcal{O}$ is mapped by $\exp$ to the geodesic γ_X i.e. $\exp(tX) = \gamma_X(t)$. This shows that the differential of $\exp_p : \mathcal{O}_p \to M$, $\mathcal{O}_p = T_p M \cap \mathcal{O}$ at the origin is the identity map of $T_p M$ under the obvious identifications.

The <u>torsion tensor</u> of a connection ∇ on M is defined by

(1.22) $\qquad T(X,Y) = \nabla_X Y - \nabla_Y X - [X,Y]$

for $X, Y \in C^\infty(TM)$. For each $p \in M$ this defines a bilinear map $T_p : T_p M \times T_p M \to T_p M$.

2. RIEMANNIAN MANIFOLDS.

In this section we will review the foundations of riemannian manifolds with special emphasis on metric properties.

A <u>riemannian manifold</u> is a manifold M with a smooth assignment g of an inner product in each tangent space of M. The <u>riemannian metric</u> g may be thought of as a section in the bundle $\text{Sym}^2(TM) \to M$ of symmetric positive definite $(2,0)$ tensors on M. In a local coordinate system $(x_1, \ldots, x_n)$

(2.1) $$g = \sum_{i,j} g_{ij}\, dx_i \otimes dx_j$$

i.e. $g\left(\frac{\partial}{\partial x_i}, \frac{\partial}{\partial x_j}\right) = g_{ij}$. We usually prefer to use to notation $\langle\,,\,\rangle$ instead of g.

Note that any submanifold M of euclidian space comes equipped in a natural way with a riemannian metric : the inner product between tangent vectors is simply the inner product taken in the ambient euclidian space.

For tangent vectors $X, Y \in T_p M$ the <u>norm</u>, $\|\cdot\|$ is given by

(2.2) $$\|X\| = \langle X, X \rangle^{\frac{1}{2}}$$

and the angle $\measuredangle$ by

(2.3) $$\langle X, Y \rangle = \|X\| \cdot \|Y\| \cos(\measuredangle(X,Y)).$$

The <u>length</u>, L of a piecewise smooth curve $\gamma : [a,b] \to M$ is given by

(2.4) $$L(\gamma) = \int_a^b \|\dot\gamma\|.$$

This will be used to define a distance function $d : M \times M \to \mathbb{R}$ on M.

First, however, we recall that for every riemannian manifold (M, g) there is a unique connection ∇ which is <u>riemannian</u> i.e.

(2.5) $$X \langle Y, Z \rangle = \langle \nabla_X Y, Z \rangle + \langle Y, \nabla_X Z \rangle$$

for all $X, Y, Z \in C^\infty(TM)$, and <u>torsion free</u> i.e.

(2.6) $$T(X,Y) = \nabla_X Y - \nabla_Y X - [X,Y]$$

for all $X, Y \in C^\infty(TM)$. This is called the <u>Levi-Civita connection</u> for (M,g).

It is determined by the identity

(2.7)
$$\langle \nabla_X Y, Z \rangle = \tfrac{1}{2}\{X\langle Y,Z\rangle - Z\langle X,Y\rangle + Y\langle Z,X\rangle$$
$$- \langle X,[Y,Z]\rangle + \langle Z,[X,Y]\rangle + \langle Y,[Z,X]\rangle\}$$

for all $X,Y,Z \in C^\infty(TM)$, or in local coordinates $(x_1,\ldots,x_n)$

(2.7')
$$\Gamma^k_{ij} = \tfrac{1}{2} \sum_\ell g^{k\ell} \left\{ \frac{\partial g_{j\ell}}{\partial x_i} + \frac{\partial g_{\ell i}}{\partial x_j} - \frac{\partial g_{ij}}{\partial x_\ell} \right\},$$

where $\{g^{k\ell}(p)\}$ is the inverse of $\{g_{k\ell}(p)\}$. The functions Γ^k_{ij} in (2.7') are called the <u>Christoffel symbols</u> of ∇.

Note that (2.5) is equivalent to saying that parallel transport along curves in M preserves the inner product.

We will now give a metric characterization of the geodesics on M. For this purpose we consider the arc length functional L of (2.4) on the space of piecewise smooth curves $\gamma : [a,b] \to M$ on M.

A <u>variation</u> of $\gamma : [a,b] \to M$ is a continous map

$$V : [a,b] \times (-\varepsilon,\varepsilon) \to M$$

for which there is a subdivision say $a = t_0 < t_1 < \ldots < t_k = b$ of $[a,b]$ such that the restriction of V to $[t_{i-1}, t_i] \times (-\varepsilon, \varepsilon)$ is smooth, $i = 1,\ldots,k$ and $V(t,0) = \gamma(t)$, $t \in [a,b]$.

The "coordinate" vector fields X,Y are the fields

$$X = V_* \circ D_1, \quad Y = V_* \circ D_2$$

along V. The curves $V(\cdot,s) : [a,b] \to M$ are denoted by γ_s. It can be helpful to think of the variation as a curve $s \to \gamma_s$ in the space of paths on M. The <u>variation field</u> Y is then the velocity field of the curve $s \to \gamma_s$.

Note that if Y is a (piecewise smooth) vector field along γ, then there is a variation V of γ with Y as variation field along γ: Simply let $V(t,s) = \exp(s \cdot Y(t))$ (cf.(1.21)).

If V is a smooth variation and γ is parametrized proportional to arc length i.e. $\|\dot\gamma\| = \ell$ then clearly $(t,s) \to \|\dot\gamma_s(t)\|$ and hence $s \to L(\gamma_s)$ is smooth for s near 0. For the derivative we have

$$\frac{d}{ds} L(\gamma_s) = \frac{d}{ds} \int_a^b \|\dot{\gamma}_s(t)\| \, dt$$

$$= \int_a^b \frac{d}{ds} <X,X>^{\frac{1}{2}} dt$$

$$= \int_a^b \frac{1}{\|X\|} <\nabla_{D_2} X, X> dt \quad , \quad \nabla \text{ riemannian}$$

$$= \int_a^b \frac{1}{\|X\|} <\nabla_{D_1} Y, X> dt \quad , \quad \nabla \text{ torsion free}$$

i.e. $\quad \dfrac{d}{ds} L(\gamma_s)(0) = \dfrac{1}{\ell} \int_a^b <\nabla_{D_1} Y, \dot{\gamma}> dt$

$$= \frac{1}{\ell} \left\{ <Y,\dot{\gamma}> \Big|_a^b - \int_a^b <Y, \nabla_{D_1} \dot{\gamma}> \right\} \; .$$

Thus for a piecewise smooth V with $\gamma_0 = \gamma$ parametriced as above we get immediately

(2.8) $\quad \dfrac{d}{ds} L(\gamma_s)(0) = \dfrac{1}{\ell} \left\{ \sum_{i=0}^{k} <Y, \dot{\gamma}^- - \dot{\gamma}^+> \Big|_{\gamma(t_i)} - \int_a^b <Y, \nabla_D \dot{\gamma}> \right\}$,

where $\dot{\gamma}^-$ and $\dot{\gamma}^+$ denote left and right limits of $\dot{\gamma}$ at the discontinuity points $\gamma(t_i)$, in particular $\dot{\gamma}^-(a) = 0$, $\dot{\gamma}^+(b) = 0$.

The formula (2.8) is referred to as the <u>first variation of arc length</u>.

As an immediate consequence of (2.8) one gets that a piecewise smooth curve γ parametrized proportional to arclength is a geodesic iff the first variation of arclength is zero for all variations of γ with fixed end points i.e. the geodesics are <u>stationary</u> points for the length functional.

The next application of (2.8) shows that the exponential map of the normal bundle $TN^\perp \to N$ of a submanifold $N \subset M$ is a radial isometry. This is a very useful generalization of what is usually referred to as the Gauss-lemma. To prove this we need a little preparation.

For a submanifold $N \subset M$ recall that the Levi-Civita connection ∇^T for the induced riemannian metric on N is given by

(2.9) $\qquad\qquad\qquad \nabla^T_X Y = (\nabla_X Y)^T$

for $X, Y \in C^\infty(TN)$. Similarly

(2.10) $$\nabla^\perp_X \eta = (\nabla_X \eta)^\perp$$

for $X \in C^\infty(TN)$, $\eta \in C^\infty(TN^\perp)$ defines a <u>riemannian connection</u> (cf. (2.5)) on the normal bundle $TN^\perp \to N$ of N in M. Here

$$TN^\perp = \{\eta \in T_p M \mid p \in N, <\eta, X> = 0, X \in T_p N\}.$$

The normal component $(\nabla_X Y)^\perp$, $X, Y \in C^\infty(TN)$ (equivalently tangent component $(\nabla_X \eta)^T$, $X \in C^\infty(TN)$, $\eta \in C^\infty(TN^\perp)$) is the <u>second fundamental tensor</u> of N.

The total space $TN^\perp$ carries a natural riemannian metric $<<,>>$: Vertical and horizontal vectors are declared to be orthogonal. For vertical vectors the "fiber metrics" is used and for horizontal vectors the "base metric" is used.

The <u>radial field</u> $\dot R$ on $TN^\perp$ is vertical field which on each fiber $T_p N^\perp$ is the gradient of $\frac{1}{2} \|\cdot\|^2$.

<u>2.11 (Gauss' Lemma)</u>. The exponential map $\exp: TN^\perp \cap \mathcal{O} \to M$ is a radial isometry i.e.

$$<\exp_*(w), \exp_*(R)>_{\exp(\eta)} = <<w, R>>_\eta$$

for all $w \in T_\eta TN^\perp$, $\eta \in TN^\perp \cap \mathcal{O}$.

<u>Proof.</u> Let $\eta(s)$ be a curve in $TN^\perp \cap \mathcal{O}$ with $\dot\eta(0) = w$ and let $c(s) = \pi \circ \eta(s)$ be the projection of η in N. We can think of η as a normal field along c. Consider the variation $V(t,s) = \exp(t \cdot \eta(s))$ of the geodesic $t \to \gamma(t) = \exp(t \cdot \eta(0))$. By the 1st variation (2.8)

(*) $$\frac{d}{ds} L(\gamma_s)(0) = \frac{1}{\|\eta\|} <\exp_*(w), \exp_*(R)>$$

since $<\dot c(0), \eta> = 0$. On the other hand

(**) $$\frac{d}{ds} L(\gamma_s)(0) = \frac{d}{ds} \|\eta(s)\|(0)$$

$$= \frac{1}{\|\eta(0)\|} <\nabla^\perp_{D_2} \eta(0), \eta(0)>, \quad \nabla^\perp \text{ riemannian}$$

$$= \frac{1}{\|\eta\|} <<w, R>>,$$

where we have used the canonical identification of $T_{\pi(\eta)} N^\perp$ with the vertical space $V_\eta \subset T_\eta TN^\perp$ together with the definition of $<<,>>$ and the definition of the covariant derivative (1.14). Combining (*) with (**) completes the proof.

For $\eta \in TN^\perp \cap O$ and $\varphi(t) = t \cdot \eta$ it is now rather straightforward to prove that

(2.12) $$L(\exp \circ \psi) \geq L(\exp \circ \varphi)$$

for any curve $\psi : [0,1] \to TN^\perp \cap O$ with $\psi(0) \in N \subset TN^\perp$ and $\psi(1) = \eta$. Strict inequality in (2.12) holds whenever there is a $t \in (0,1]$ such that the component of $\dot\psi(t)$ perpendicular to $R_{\psi(t)}$ is not in the kernel of $\exp_*$.

Since $\exp : TN^\perp \cap O \to M$ has maximal rank along the zero section $N \subset TN^\perp$ (cf. p.1.8) there is by the inverse function-theorem an open (starshaped) neighborhood D of N in $TN^\perp \cap O$ such that $\exp : D \to \exp(D) = U$ is a diffeomorphism. U is called a <u>tubular neighborhood</u> of N in M.

For points $q = \exp_p(\eta) \in U$ it follows that the geodesic $t \to \exp((1-t)\eta)$ perpendicular to N is the shortest curve from q to N. The length of this curve is clearly $\|\eta\|$.

Now define the <u>distance</u> $d(p,q)$ between points $p, q \in M$ by the formula

(2.13) $$d(p,q) = \inf L(\gamma)$$

where the infimum is taken over all piecewise smooth curves γ from p to q.

It follows directly from the above considerations (with N a point) that d is a metric on each path component of M. Moreover the metric topology is the same as the manifold topology.

We recall without proof the fundamental

<u>2.14 (Hopf-Rinow.)</u> For a connected riemannian manifold $(M, <,>)$ the following are equivalent:

(i) (M,d) is a complete metric space

(ii) There is a $p \in M$ with $O_p = T_p M$

(iii) $O = TM$.

If one (hence any) of the above holds then any $p, q \in M$ can be joined by a minimal geodesic.

Unless otherwise stated all manifolds considered from now on will be assumed to be complete.

Our main concern is to examine the intimate relations between topology, curvature and metric properties of complete riemannian manifolds.

The following identities for the curvature tensor (1.19) of the Levi-Civita connection (2.7) are obtained by straightforward essentially algebraic manipulations

(2.15) (i) $R(X,Y)Z = -R(Y,X)Z$

(ii) $R(X,Y)Z + R(Y,Z)X + R(Z,X)Y = 0$

(iii) $<R(X,Y)Z,U> = -<R(X,Y)U,Z>$

(iv) $<R(X,Y)Z,U> = <R(Z,U)X,Y>$

for all $X,Y,Z,U \in C^\infty(TM)$.

The sectional curvature $K(\sigma)$ of a twoplane $\sigma = \text{span }\{X,Y\}$ is defined as

(2.16) $$K(\sigma) = \frac{<R(X,Y)Y,X>}{\|X\|^2 \|Y\|^2 - <X,Y>^2}.$$

If in particular $\sigma = \text{span}(U,V)$ for U and V orthonormal then $K(\sigma) = <R(U,V)V,U>$.

The sectional curvature is obviously determined by the curvature tensor, but the converse is true as well.

A riemannian manifold with constant sectional curvature K is called a space form. A space form is said to be hyperbolic if $K<0$, flat if $K=0$, and elliptic if $K>0$. The curvature tensor of a space form with curvature k is given by $R(X,Y)Z = k\{<Z,Y>X - <Z,X>Y\}$ for $X,Y,Z \in C^\infty(TM)$.

2.17 Examples.

(i) The sphere $S^n(r) \subset R^{n+1}$ with radius $r>0$ has constant curvature $K = \frac{1}{r^2}$.

(ii) The euclidian space R^n with its usual riemannian metric has constant curvature $K=0$.

(iii) For $\kappa<0$ consider the riemannian metric $g = \varphi_\kappa \cdot g$ on the open disc $U_\kappa = D\left(\frac{1}{\sqrt{|\kappa|}}\right) \subset R^n$ of radius $\frac{1}{\sqrt{|\kappa|}}$. Here g is the euclidian metric and $\varphi_\kappa : U_\kappa \to R_+$ is defined by $\varphi_\kappa(x) = 4(1 + \kappa \|x\|^2)^{-2}$. U_κ with the metric g_κ has constant curvature κ. If $\kappa = -1$ this is called the __Poincaré model__ for the hyperbolic n-space.

(iv) For $\kappa \geq 0$ and $U_\kappa = R^n$ the metric g_κ in (iii) has constant curvature κ. For $\kappa = \frac{1}{r^2}$ this metric is the pull back by the stereographic projection of the standard metric on the n-sphere of radius r.

Riemannian metrics related as in the examples (iii), (iv) above are called __conformally equivalent__.

__The simply connected space forms__ in (2.17) play a fundamental role as model spaces in riemannian geometry. This will be apparent in our treatment of comparison theory.

Other important notions of curvature are obtained by averaging sectional curvatures.

__The Ricci-tensor__ $c_1 R$ is given by

(2.18) $$c_1 R(X_p, Y_p) = \text{trace}(U_p \to R(U_p, X_p) Y_p)$$

for all $X_p, Y_p \in T_p M$. The __Ricci curvature__ $\text{Ric}(\ell)$ of the line $\ell = \text{span}\{X\}$ is defined as

(2.19) $$\text{Ric}(\ell) = \frac{c_1 R(X,X)}{\|X\|^2}$$

Notice that if $\{X, U_1, \ldots, U_{n-1}\}$ is an orthonormal basis for $T_p M$ then

$$\text{Ric}(\ell) = \sum_{i=1}^{n-1} K(\sigma_i),$$

where $\sigma_i = \text{span}\{X, U_i\}$.

The __scalar curvature__ S is the trace of the Ricci tensor i.e.

(2.20) $$S(p) = \sum_{r=1}^{n} c_1 R(U_i, U_i),$$

for an orthonormal basis $\{U_1, \ldots, U_n\}$ for $T_p M$. Clearly

$$S(p) = 2 \sum_{1 \leq i < j \leq n} K(\sigma_{ij}),$$

where $\sigma_{ij} = \text{span}\{U_i, U_j\}$.

All these curvatures agree (essentially) for $n = 2$. Their geometric significance is illuminated in the calculus of variations and in comparison theorems.

A two-parameter variation of $\gamma : [a,b] \to M$ is a continuous map

$$W : [a,b] \times (-\varepsilon_1, \varepsilon_1) \times (-\varepsilon_2, \varepsilon_2) \to M$$

with $W(t,0,0) = \gamma(t)$, $t \in [a,b]$ such that there is a subdivision $a = t_0 < t_1 < \ldots < t_k = b$ of $[a,b]$ for which

$$W \,|\, [t_{i-1}, t_i] \times (-\varepsilon_1, \varepsilon_1) \times (-\varepsilon_2, \varepsilon_2)$$

is smooth.

As in the case of a 1-parameter variation we let $\gamma_{s_1, s_2} = W(\cdot, s_1, s_2)$ and

$$X = W_* \circ D \;,\; Y_1 = W_* \circ D_1 \;,\; Y_2 = W_* \circ D_2 \;,$$

where D, D_1 and D_2 are the standard coordinate fields on $[a,b] \times (-\varepsilon_1, \varepsilon_1) \times (-\varepsilon_2, \varepsilon_2)$.

We are interested in the second variation of arclength only if the first variation is zero i.e. at geodesics.

Now if W is a two-parameter variation of the geodesic $c : [a,b] \to M$ with $\|\dot{c}\| = \ell \neq 0$ then $(s_1, s_2) \to L(c_{s_1, s_2})$ is smooth near $(0,0)$ and

$$(2.21) \qquad \frac{\partial^2}{\partial s_1 \partial s_2} L(c_{s_1, s_2})(0,0) = \frac{1}{\ell} \{ <\nabla_{D_1} Y_2, X> \big|_a^b + I(Y_1^\perp, Y_2^\perp) \}$$

where

$$(2.22) \qquad I(Y_1^\perp, Y_2^\perp) = \int_a^b <\nabla_D Y_1^\perp, \nabla_D Y_2^\perp> - <R(Y_1^\perp, X)X, Y_2^\perp>$$

is the socalled <u>index form</u> defined on the space of vector field along c orthogonal to $\dot{c}$. The formula (2.21) is referred to as the <u>second variation of arclength</u>. If W is a smooth 2-parameter variation, we get from the proof of (2.8), that

$$\frac{\partial}{\partial s_2} L(c_{s_1, s_2}) = \int_a^b \frac{<\nabla_D Y_2, X>}{\|X\|} dt$$

and hence

$$\frac{\partial^2}{\partial s_1 \partial s_2} L(c_{s_1,s_2}) = \int_a^b \frac{<\nabla_{D_1}\nabla_D Y_2, X> + <\nabla_D Y_2, \nabla_{D_1} X>}{\|X\|} dt$$

$$- \int_a^b \frac{<\nabla_D Y_2, X> <\nabla_{D_1} X, X>}{\|X\|^3} dt, \text{ by (2.5)}$$

$$= \int_a^b \frac{<R(Y_1,X)Y_2, X> + <\nabla_D \nabla_{D_1} Y_2, X> + <\nabla_D Y_2, \nabla_D Y_1>}{\|X\|} dt$$

$$- \int_a^b \frac{<\nabla_D Y_2, X> <\nabla_D Y_1, X>}{\|X\|^3} dt, \text{ by (1.19) and (2.6)}$$

thus at $(s_1, s_2) = (0,0)$

(2.23) $$\frac{\partial^2}{\partial s_1 \partial s_2} L = \frac{1}{\ell} \Big\{ <\nabla_{D_1} Y_2, X>\big|_a^b$$

$$+ \int_a^b <\nabla_D Y_1, \nabla_D Y_2> - <R(Y_1,X)X, Y_2>$$

$$- <\nabla_D Y_2, \frac{X}{\|X\|}> <\nabla_D Y_1, \frac{X}{\|X\|}> \Big\}.$$

Now $Y_1 = Y_1^\perp + <Y_1, \frac{X}{\|X\|}> \frac{X}{\|X\|}$ and hence $\nabla_D Y_1 = \nabla_D Y_1^\perp + <\nabla_D Y_1, \frac{X}{\|X\|}> \frac{X}{\|X\|}$ since $\nabla_D X = 0$. Then (2.21) follows from (2.23) because $<\nabla_D Y_1^\perp, X> = 0$.

Note that as for 1-parameter variations a two parameter variation gives rise to two variation fields Y_1 and Y_2 along c and vice versa (use the exponential map).

A smooth vector field Y along a geodesic c is called a <u>Jacobi field</u> if

(2.24) $$\nabla_D \nabla_D Y + R(Y, \dot{c})\dot{c} = 0.$$

This is a linear second order equation (express Y in an orthonormal basis of parallel fields along c) i.e. for any $u, v \in T_{c(t_0)} M$ there is a unique Jacobi field Y along c with $Y(t_0) = u$ and $\nabla_D Y(t_0) = v$. Geometrically any Jacobi field is the variation field for a variation through geodesics and vice versa. Specifically if $\alpha(s)$ is a curve with $\dot{\alpha}(0) = u$ and X is a vector field along α with $X(0) = \dot{c}(t_0)$ and $\nabla_u X = v$, then $V(t,s) = \exp(t \cdot X(s))$ is a variation of c with the above Jacobi field Y as variation field along c.

A <u>focal point</u> of a submanifold $N \subset M$ is a critical value of the normal exponential map $\exp : TN^\perp \to M$. Specifically $\exp(t_0 \eta)$, $\eta \in TN^\perp$ is a focal point of N along the geodesic $c(t) = \exp(t\eta)$ iff there is a $W \in T_{t_0 \eta} TN^\perp$ with $\exp_*(W) = 0$. Such a W determines in an obvious way a Jacobi field along c which is tangent to N at $c(0)$ (cf. proof of (2.11)) and zero at $c(t_0)$. The <u>multiplicity</u> of a focal point $c(t_0)$ is by definition the dimension of $\ker(\exp_{*t_0 \eta})$.

If $N = \{p\}$ a focal point $c(t_0)$ is called a <u>conjugate</u> point to $p = c(0)$ along c. The multiplicity of a conjugate point $c(t_0)$ is clearly the dimension of the space of Jacobi fields along c which vanish at 0 and at t_0. This space is also the null space for the index form I (cf. (2.22)) restricted to the space of piecewice smooth vector fields along c which vanish at 0 and at t_0. Note that by (2.21) this is the second variation of arc length for variations fixing the end points $c(0)$ and $c(t_0)$. Then index I on this space i.e. the maximal dimension of subspaces on which I is negative definite is called the <u>index</u> of $c : [0, t_0] \to M$.

The celebrated <u>index theorem</u> of M. Morse states that along any geodesic $c : [a,b] \to M$ there are at most finitely many conjugate points to $c(a)$ and the index of c is the number of conjugate points to $c(a)$ in $[a,b]$ counted with multiplicities. This is of fundamental importance in Morse theory (c.f. $[M_1]$).

We conclude this section by showing that Jacobi fields in some sense minimize the index form I. This is used in the proof of the Morse index theorem (which we will not give here) as well as in the proof of the Rauch comparison theorems.

2.25 Index lemma I. Let $c : [a,b] \to M$ be a geodesic such that no $c(t)$ is conjugate to $c(a)$ along c. For any piecewise smooth vector field X along c with $X \perp c$ and $X(a) = 0$ let Y be the unique Jacobi field with $Y(a) = 0$ and $Y(b) = X(b)$. Then $I(Y,Y) \leq I(X,X)$ and equality holds iff $X = Y$.

We will abuse language by saying that $c(t_0)$ is a "<u>focal point</u>" to $c(a)$ along c if it is a focal point along c of the unique <u>locally defined</u>

codimension 1 submanifold $N = \exp(\dot{c}(a)^\perp)$.

2.26 Index lemma II. Let $c : [a,b] \to M$ be a geodesic such that no $c(t)$ is a "focal point" to $c(a)$ along c. For any piecewise smooth vector field X along c with $X \perp \dot{c}$ let Y be the unique Jacobi field with $\nabla_D Y(a) = 0$ and $Y(b) = X(b)$. Then $I(Y,Y) \leq I(X,X)$ and equality holds iff $X = Y$.

The proofs of (2.25) and (2.26) are very similar. We will focus on the

Proof of 2.26. Note that $Z = X - Y$ is orthogonal to c and $Z(b) = 0$. Consider the variation $V(t,s) = \exp_{c(t)} s \cdot Z(t)$. Then by the second variation (2.21) we get $I(Z,Z) = \frac{\partial^2}{\partial s^2} L(c_s)$. Since $c[a,b]$ contains no "focal point" we can lift $V(t,s)$ to the normal bundle $TN^\perp$ for s near 0. By the corollary (2.12) to the Gauss-lemma we conclude that $\frac{\partial^2}{\partial s^2} L(c_s) \geq 0$ i.e.

$$0 \leq I(Z,Z) = I(X,X) + I(Y,Y) - 2 I(Y,X).$$

However,

$$I(Y,X) = \sum \int_{a_i}^{b_i} \frac{d}{dt} <\nabla_D Y, X> - <\nabla_D \nabla_D Y, X> - <R(Y,\dot{c})\dot{c}, X>$$

$$= <\nabla_D Y, X> \Big|_a^b$$

$$= <\nabla_D Y, Y> \Big|_a^b$$

$$= I(Y,Y)$$

and hence $I(Y,Y) \leq I(X,X)$. Moreover since I is positive semi definite on the space of vector fields orthogonal to c and zero at b it follows that $I(Z,Z) = 0$ iff $I(U,Z) = 0$ for all U in this space of vector fields along c. From the second variation of arc length this implies that Z is a Jacobi field i.e. $Z = 0$ since $c(b)$ is not a "focal point" to $c(a)$ along c. This completes the equality discussion.

3. COMPARISON THEOREMS.

The main idea behind comparison theory is to be able to estimate basic geometric quantities such as <u>distance functions and volume functions</u> for general riemannian manifolds with curvature bounds, in terms of corresponding quantities for space forms.

We begin with the fundamental so called <u>Rauch comparison theorems</u>.

To fix notation consider riemannian manifolds $M^{n+k}, \bar{M}^n$, and <u>normal geodesics</u> $\gamma, \bar{\gamma} : [0,\ell] \to M, \bar{M}$ i.e. $\|\dot{\gamma}\| = \|\dot{\bar{\gamma}}\| = 1$. We say that the <u>sectional curvature along γ is bounded below by the sectional curvature along $\bar{\gamma}$</u> when for each $t \in [0,\ell]$ and any $X \in T_{\gamma(t)}M$, $\bar{X} \in T_{\bar{\gamma}(t)}\bar{M}$ the sectional curvatures of the planes $\sigma, \bar{\sigma}$ spanned by $X, \dot{\gamma}$ and $\bar{X}, \dot{\bar{\gamma}}$ satisfy $K(\sigma) \geq K(\bar{\sigma})$.

3.1 Rauch I. Let $M, \bar{M}$ and $\gamma, \bar{\gamma} : [0,\ell] \to M, \bar{M}$ be as above and assume that the sectional curvature along γ is bounded below by the sectional curvature along $\bar{\gamma}$. Assume moreover that no $\gamma(t)$, $t \in [0,\ell]$ is conjugate to $\gamma(0)$ along γ. Now if $Y, \bar{Y}$ are Jacobi fields along $\gamma, \bar{\gamma}$ such that

$$Y(0) = c\dot{\gamma}(0) \quad , \quad \bar{Y} = c \cdot \dot{\bar{\gamma}}(0)$$
$$\|Y'(0)\| = \|\bar{Y}'(0)\|$$
$$\langle Y'(0), \dot{\gamma}(0) \rangle = \langle \bar{Y}'(0), \dot{\bar{\gamma}}(0) \rangle$$

then

$$\|Y(t)\| \leq \|\bar{Y}(t)\|$$

for all $t \in [0,\ell]$.

For simplicity we will (as above) use Y' to denote the covariant derivative of Y.

The following important analogue of Rauch I is due to M. Berger.

3.2 Rauch II. Let $\gamma, \bar{\gamma} : [0,\ell] \to M, \bar{M}$ and the sectional curvature along $\gamma, \bar{\gamma}$ be as in (3.1). Suppose moreover that no $\gamma(t)$, $t \in [0,\ell]$ is a "focal point" (cf. p. 2.10) to $\gamma(0)$ along γ. If now $Y, \bar{Y}$ are Jacobi fields along $\gamma, \bar{\gamma}$ such that

$$Y'(0) = c\,\dot{\gamma}(0)\;,\quad \bar{Y}'(0) = c\cdot\dot{\bar{\gamma}}(0)$$

$$\|Y(0)\| = \|\bar{Y}(0)\|$$

$$<Y(0),\dot{\gamma}(0)> \;=\; <\bar{Y}(0),\dot{\bar{\gamma}}(0)>$$

then

$$\|Y(t)\| \le \|\bar{Y}(t)\|$$

for all $t \in [0,\ell]$.

The proofs of (3.1) and (3.2) are based on (2.25) and (2.26) respectively. We confine ourselves to

<u>Proof of 3.2.</u> By splitting $Y,\bar{Y}$ into tangential and orthogonal components we see that it suffices to prove (3.2) for $Y,\bar{Y}$ perpendicular to $\gamma,\bar{\gamma}$ i.e. $c = 0$. Now fix t_0 and suppose $\bar{Y}$ has no zero's in $[0,t_0]$. To complete the proof it is clearly enough to show that $\|Y(t_0)\| \le \|\bar{Y}(t_0)\|$. Observe that since γ has no "focal points" we may assume that Y has no zero's (or else is identically zero). Choose a linear isometric imbedding

$$\iota_0 : T_{\bar{\gamma}(0)}\bar{M} \to T_{\gamma(0)}M \text{ that maps } \dot{\bar{\gamma}}(0),\bar{Y}_{t_0} \text{ to } \dot{\gamma}(0),\frac{\|\bar{Y}(t_0)\|}{\|Y(t_0)\|}\cdot Y_{t_0}(0),$$

where $Y_{t_0}(t)$ denotes the unique parallel field along γ with $Y_{t_0}(t_0) = Y(t_0)$. Extend ι_0 to an isometry

$$\iota : C^\infty(\bar{\gamma}^\perp) \to C^\infty(\gamma^\perp)$$

by setting

$$\iota(X)(t) = \iota_0((X_t)(0))_0(t)$$

for any vectorfield $X \in C^\infty(\gamma^\perp)$ along γ perpendicular to γ. Then

$$\frac{d}{dt}\|\bar{Y}\|^2(t_0) = 2<\bar{Y}',\bar{Y}>(t_0) = 2<\bar{Y}',\bar{Y}>\Big|_0^{t_0}$$

$$= 2\,I(\bar{Y},\bar{Y})$$

$$\ge 2\,I(\iota\bar{Y},\iota\bar{Y}) \qquad \text{by curvature assumption}$$

$$\ge 2\,\frac{\|\bar{Y}(t_0)\|^2}{\|Y(t_0)\|^2} \cdot I(Y,Y) \quad \text{by} \quad (2.26)$$

$$= \frac{\|\bar{Y}(t_0)\|^2}{\|Y(t_0)\|^2}\,\frac{d}{dt}\|Y\|^2(t_0)$$

i.e.

$$\frac{d}{dt}\log\|\bar{Y}\|^2(t) \ge \frac{d}{dt}\log\|Y\|^2(t)$$

holds for all $0 \le t \le t_0$. By integration we conclude $\|\bar{Y}(t_0)\| \ge \|Y(t_0)\|$.

Note from the proof that equality at $t = \ell$ in (3.2) implies equality for all $t \in [0,\ell]$. Moreover $\iota \bar{Y} = Y$ and the sectional curvature along γ spanned by $\dot{\gamma}$ and Y agrees with the sectional curvature along $\bar{\gamma}$ spanned by $\dot{\bar{\gamma}}$ and $\bar{Y}$.

The Rauch comparison theorem allow to compare lengths of curves in riemannian manifolds $M^{n+k}, \bar{M}^n$ for which $K_M \ge K_{\bar{M}}$ i.e. for all planes $\sigma, \bar{\sigma}$ in $M, \bar{M}$ the sectional curvatures satisfy $K(\sigma) \ge K(\bar{\sigma})$.

Specifically let $p, \bar{p} \in M, \bar{M}$ and let $I : T_{\bar{p}}\bar{M} \to T_p M$ be a linear isometric imbedding. Suppose $\exp_p |B_r(0)$ is non-singular and $\exp_{\bar{p}} |B_r(0)$ is an imbedding. Then

(3.3) $$L(c) \ge L(\bar{c}),$$

follows from Rauch I for any curve $c = \exp_p \circ I \circ \exp_{\bar{p}}^{-1} \circ \bar{c}$ with $\bar{c} : [0,\ell] \to \exp_{\bar{p}} B_r(0)$.

The following analogous corollary of Rauch II is crucial for our proof of Toponogov's triangle comparison theorem.

(3.4). Let $M^{n+k}, \bar{M}^n$ be as above and let $E, \bar{E}$ be parallel fields along normal geodesics $\gamma, \bar{\gamma} : [0,\ell] \to M, \bar{M}$ satisfying

$$\|E\| = \|\bar{E}\|, \quad <E,\dot{\gamma}> = <\bar{E},\dot{\bar{\gamma}}>.$$

Suppose $c = \exp \circ fE$, $\bar{c} = \exp \circ f\bar{E}$ for some $f : [0,\ell] \to R$. Then

$$L(c) \le L(\bar{c})$$

if for every $t \in [0,\ell]$ the geodesic $s \to \exp(s f(t) E(t))$ has no "focal points" to $\gamma(t)$.

Proof. Consider the variations

$$V(s,t) = \exp(s f(t) E(t)), \quad \bar{V}(s,t) = \exp(s f(t) \bar{E}(t))$$

and observe that Rauch II applies directly to the Jacobi fields

$$Y = V_* \circ \frac{\partial}{\partial t}, \quad \bar{Y} = \bar{V}_* \circ \frac{\partial}{\partial t}.$$

This yields $\|\dot{c}(t)\| \le \|\dot{\bar{c}}(t)\|$ for all $t \in [0,\ell]$.

The next theorem is a global comparison theorem. In order to state it we need some definitions.

A <u>geodesic triangle</u> $(\gamma_1,\gamma_2,\gamma_3)$ in a riemannian manifold M is a configuration of three normal geodesics $\gamma_i : [0,\ell_i] \to M$ (the <u>sides</u>) such that

$$\gamma_i(\ell_i) = \gamma_{i+1}(0) ,$$

$$\ell_i + \ell_{i+1} \geq \ell_{i+2}$$

indices taken modulo 3. The points $p_i = \gamma_{i+2}(0)$ are called the <u>vertices</u> of the triangle and $\alpha_i = ((-\dot{\gamma}_{i+1}(\ell_{i+1}), \dot{\gamma}_{i+2}(0))$ the corresponding <u>angles</u>.

A <u>geodesic hinge</u> $(\gamma_1,\gamma_2,\alpha)$ in a riemannian manifold is a configuration of two normal geodesics $\gamma_i : [0,\ell] \to M$ which meet at $p = \gamma_1(\ell_1) = \gamma_2(0)$ at an angle $\alpha = ((\dot{\gamma}_1(\ell_1), -\dot{\gamma}_2(0))$.

It is useful to state (and prove simultaniously) two equivalent versions of the so called <u>Aleksandrow-Toponogov</u> triangle comparison theorem. Recall the classical result that any two simply connected complete riemannian manifolds with constant (sectional) curvature κ are isometric. Let M_κ^n be the simply connected n-dimensional space form of constant curvature κ.

<u>3.5 Toponogov.</u> Let M be a complete riemannian manifold with sectional curvature $K \geq \kappa$.

(A) Let $(\gamma_1,\gamma_2,\gamma_3)$ be a geodesic triangle in M. Suppose γ_1,γ_3 are minimal and if $\kappa > 0$ suppose $L(\gamma_2) = \ell_2 \leq \pi/\sqrt{\kappa}$. Then there exists a geodesic triangle $(\bar{\gamma}_1,\bar{\gamma}_2,\bar{\gamma}_3)$ in M_κ^2 such that $L(\gamma_i) = L(\bar{\gamma}_i)$ and $\bar{\alpha}_1 \leq \alpha_1, \bar{\alpha}_3 \leq \alpha_3$. Except in case $\kappa > 0$ and $L(\gamma_i) = \pi/\sqrt{\kappa}$ for some i, the triangle $(\bar{\gamma}_1,\bar{\gamma}_2,\bar{\gamma}_3)$ is uniquely determined.

(B) Let $(\gamma_1,\gamma_2,\alpha)$ be a geodesic hinge in M. Suppose γ_1 is minimal and if $\kappa > 0$, suppose $L(\gamma_2) \leq \pi/\sqrt{\kappa}$. Let $(\bar{\gamma}_1,\bar{\gamma}_2,\alpha)$ be a geodesic hinge in M_κ^2 such that $L(\gamma_i) = L(\bar{\gamma}_i)$. Then

$$d(\gamma_1(0), \gamma_2(\ell_2)) \leq d(\bar{\gamma}_1(0), \bar{\gamma}_2(\ell_2)) .$$

The following elementary facts will be used over and over in the proof of (3.5).

<u>(3.6).</u> In M_κ^2 fix $\bar{\gamma}_1$ and let $f(\alpha) = d(\bar{\gamma}_1(0), \bar{\gamma}_2(\ell_2))$ where $(\bar{\gamma}_1,\bar{\gamma}_2,\alpha)$ is a hinge in M_κ^2. Then $\alpha \to f(\alpha)$ is strictly increasing on $[0,\pi]$ except when $\kappa > 0$ and $\ell_i = \pi/\sqrt{\kappa}$ for some i in which case f is constant. Note that $f(0) = |\ell_1 - \ell_2|$ and $f(\pi) = \min\left\{\ell_1 + \ell_2, (2\pi/\sqrt{\kappa}) - \ell_1 - \ell_2\right\}$.

Here, and in the following, conditions involving $\sqrt{\kappa}$ should be ignored if $\kappa \leq 0$.

Proof. The exceptional cases are obvious. In all other cases there is a unique minimal geodesic $\bar{c}_\alpha$ from $\bar{\gamma}(0)$ to $\bar{\gamma}_2(\ell_2)$ and f is smooth. Our claim is then an obvious consequence of the 1st variation formula (2.8) applied to the variation $\alpha \to \bar{c}_\alpha$.

(3.7). In M_κ^2 a triangle $(\bar{\gamma}_1, \bar{\gamma}_2, \bar{\gamma}_3)$ with side lengths $\ell_i \leq \pi/\sqrt{\kappa}$ is uniquely determined by (ℓ_1, ℓ_2, ℓ_3) unless $\ell_i = \pi/\sqrt{\kappa}$ for some i.

Proof. For fixed ℓ_1, ℓ_2 it follows from (3.6) that ℓ_3 uniquely determines α_3 i.e. $(\bar{\gamma}_1, \bar{\gamma}_2, \bar{\gamma}_3)$ is determined by the hinge $(\bar{\gamma}_1, \bar{\gamma}_2, \alpha_3)$. The claim now follows by homogeneity of M_κ^2.

Proof of 3.5.

Step 1. The statements (A) and (B) are equivalent.

This follows directly from (3.6) and (3.7).

Step. 2. (B) holds for "thin" hinges.

Let $(\gamma_1, \gamma_2, \alpha)$ be a hinge in M and consider the parallel field E along γ_1 determined by $E(\ell_1) = \dot{\gamma}_2(0)$. We say that $(\gamma_1, \gamma_2, \alpha)$ is "thin" if for the corresponding $(\bar{\gamma}_1, \bar{\gamma}_2, \alpha)$ and $\bar{E}$ in M_κ^2 the curve $\bar{c}(t) = \exp(f(t)\bar{E}(t))$ is a minimal curve from $\bar{\gamma}_1(0)$ to $\bar{\gamma}_2(\ell_2)$ for some $f : [0, \ell_1] \to R$, and moreover for each $t \in [0, \ell_1]$ there are no "focal points" to $\gamma_1(t)$ along the geodesic $s \to \exp(s \cdot f(t)E(t))$ in M. By the corollary (3.4) to Rauch II we then have

$$L(c) \leq L(\bar{c}) = d(\bar{\gamma}_1(0), \bar{\gamma}_2(\ell_2)),$$

where $c = \exp \circ f \cdot E$ is a curve from $\gamma_1(0)$ to $\gamma_2(\ell_2)$. Since by definition (2.13)

$$d(\gamma_1(0), \gamma_2(\ell_2)) \leq L(c)$$

we have verified (B) for "thin" hinges.

Step 3. (B) holds for "generic" hinges.

We will say that $(\gamma_1, \gamma_2, \alpha)$ is "generic" if $\ell_1, \ell_2 < \pi/\sqrt{\kappa}$ and $\alpha < \pi$. It is clearly sufficient to prove (B) for hinges $(\gamma_1, \gamma_2, \alpha)$ which satisfy

$$\ell_2 < {}^\pi/\sqrt{\kappa} \quad \text{and} \quad \max d\,(\gamma_1(0)\,,\gamma_2(t)) = d_2 < {}^\pi/\sqrt{\kappa} \;.$$

Now consider the compact set of all minimal geodesics from $\gamma_1(0)$ to $\gamma_2(t)$, $t \in [0,\ell_2]$. An easy compactness argument now shows that we can choose a subdivision

$$0 = t_0 < t_1 < \ldots < t_k = \ell_2 \,, \quad |t_{i+1} - t_i| < {}^\pi/\sqrt{\kappa} - d_2 \,,$$

and minimal geodesics γ_{3i} from $\gamma_1(0)$ to $\gamma_2(t_i)$ beginning with $\gamma_{30} = \gamma_1$ such that all the hinges

$$(\gamma_{3i}, \gamma_2 \mid [t_i, t_{i+1}], \alpha_1)$$

and

$$(\gamma_{3i}, \gamma_2^{-1} \mid [t_i, t_{i-1}], \beta_i)$$

are "thin". Here obviously $\alpha = \alpha_0$. Note also that to prove (B) we can assume $\ell_2 \leq \ell_1 + d(\gamma_1(0), \gamma_2(\ell_2))$. In this case all triangle inequalities hold for all the triangles $(\gamma_{3i}, \gamma_2 \mid [t_i, t_j], \gamma_{3j}^{-1})$, $i < j$. We now proceed by "forward and backward" induction: Since (B) holds for the hinges $(\gamma_{30}, \gamma_2 \mid [t_0, t_1], \alpha_0)$ and $(\gamma_{31}, \gamma_2^{-1} \mid [t_1, t_0], \beta_1)$ we get that (A) holds for the triangle $(\gamma_{30}, \gamma_2 \mid [t_0, t_1], \gamma_{31}^{-1})$ (cf. step 1). Since (B) holds for the hinge $(\gamma_{30}, \gamma_2 \mid [t_1, t_2], \alpha_1)$ we get from this using (3.6) twice that (B) holds for the hinge $(\gamma_{30}, \gamma_2 \mid [t_0, t_2], \alpha_0)$. Similarly going backwards i.e. beginning with the hinges $(\gamma_{32}, \gamma_2^{-1} \mid [t_2, t_1], \beta_2)$ and $(\gamma_{31}, \gamma_2 \mid [t_1, t_2], \alpha_1)$ we obtain that (B) holds for the hinge $(\gamma_{32}, \gamma_2^{-1} \mid [t_2, t_0], \beta_2)$. Again by step 1 we get that (A) holds for the triangle $(\gamma_{30}, \gamma_2 \mid [t_0, t_2], \gamma_{32}^{-1})$. Proceeding back and forth like this finitely many times proves the claim.

Step 4. (B) holds for all hinges.

Use step 3 and two easy limit arguments (first on α then on ℓ_i).

There is a <u>rigidity part</u> to (3.5 B) which follows directly from our proof by careful examination of the steps.

3.8 Rigidity theorem. In the situation (3.5 B) suppose $\ell_2 < {}^\pi/\sqrt{\kappa}$, $0 < \alpha < \pi$ and $d(\gamma_1(0), \gamma_2(\ell_2)) = d(\bar\gamma_1(0), \bar\gamma_2(\ell_2))$. Then there is an isometric and totally geodesic imbedding of the triangular surface in M_κ^2 determined by $(\bar\gamma_1, \bar\gamma_2, \alpha)$ to M, which maps $(\bar\gamma_1, \bar\gamma_2, \alpha)$ to $(\gamma_1, \gamma_2, \alpha)$. Moreover the image of the minimal geodesics in M_κ^2 from $\bar{\gamma}_1(0)$ to $\bar{\gamma}_2(t)$ are minimal geodesics in M from $\gamma_1(0)$ to $\gamma_2(t)$.

Let us point out that the "obvious analogue" to (3.5) where the lower curvature bound is replaced by an upper curvature bound is wrong! This is related to the interesting phenomenon of collapse, a topic which belongs naturally to the context of section 5. Here we mention only that there are even homogeneous metrics on the 3-sphere S^3 (the Berger-spheres) with positive sectional curvature $K \leq 1$ and closed geodesics of length $< 2\pi$. On the other hand it follows immediately from the corollary (3.3) of Rauch I, that

3.9 "Inverse Toponogov". Let M be a complete riemannian manifold with sectional curvature $K \leq \kappa$. Let $(\gamma_1,\gamma_2,\gamma_3)$ be a geodesic triangle in M with minimal sides and $\sum \ell_i < 2\pi/\sqrt{\kappa}$. Suppose moreover that $(\gamma_1,\gamma_2,\gamma_3)$ is contained in a normal ball $B_r(p_3) \subset M$ i.e. $\exp_{p_3}|B_r(0)$ is an imbedding. Then there is a triangle $(\bar{\gamma}_1,\bar{\gamma}_2,\bar{\gamma}_3)$ in M_κ^2 with $L(\gamma_i) = L(\bar{\gamma}_i)$ and $\alpha_3 \leq \bar{\alpha}_3$.

It is natural to interpret Toponogov's theorem (B) as a statement about the distance function $d(\gamma_1(0), \cdot)$.

We now turn to the volume function. First some preparation.

On any riemannian manifold $(M, <,>)$ there is a canonical measure, $d\mu$, called the riemannian measure (or volume element). In local coordinates $(x_1,\ldots,x_n)$ on $U \subset M$ this measure is given by

(3.10) $$d\mu = \sqrt{\det\{g_{ij}\}}\, d\lambda,$$

where $g_{ij} = <\frac{\partial}{\partial x_i}, \frac{\partial}{\partial x_j}>$ and $d\lambda = dx_1,\ldots,dx_n$ is the standard Lebesque measure on R^n. Note that $\sqrt{\det\{g_{ij}(p)\}} = |\frac{\partial}{\partial x_1}(p) \ldots \frac{\partial}{\partial x_n}(p)|$ where $|v_1 \ldots v_n|$, $v_i \in T_pM$ is the volume of the n-dimensional parallelepipedum in T_pM spanned by $v_1,\ldots,v_n$.

The volume of a measureable set $A \subset M$ is then

(3.11) $$\text{vol}(A) = \int_M 1_A \cdot d\mu$$

and in particular $$\text{vol}(M) = \int_M d\mu.$$

Recall that the tangent cut locus of $p \in M$ is the set

(3.12) $$C_p = \{t_v \cdot v \in T_p M \mid \|v\| = 1\} ,$$

where $t_v = \sup\{t > 0 \mid d(p, \exp t \cdot v) = t\}$ is the __cut value__ of v.
The __cut locus__ of p is the set of __cut points__, $\exp(t_v \cdot v) = \gamma_v(t_v)$.

(3.13) $$C(p) = \exp(C_p) .$$

The following gives a useful characterization of cut points.

(3.14) Let $\gamma : [0,\infty) \to M$ be a normal geodesic in M. Then $\gamma(t_0)$ is a cut point of $p = \gamma(0)$ if and only if either

(i) $\gamma(t_0)$ is conjugate to p along γ

or

(ii) there is a minimal geodesic σ from p to $\gamma(t_0)$ different from γ, and neither holds for any $t < t_0$.

__Proof.__ Let $\varepsilon_i > 0$ be a sequence with $\varepsilon_i \to 0$. Let $\sigma_i \neq \gamma$ be a sequence of minimal geodesics from p to $\gamma(t_0 + \varepsilon_i)$ with $\|\dot\sigma_i(0)\| = 1$. By compactness of the unit sphere in $T_p M$ we may assume that $\dot\sigma_i(0) \to \dot\sigma(0)$, where σ is a minimal geodesic from p to $\gamma(t_0)$. Either $\sigma \neq \gamma$ or by the inverse function theorem $t_0 \dot\gamma(0)$ is a critical point of $\exp_p$. Conversely, if $\gamma(t_0)$ is conjugate to $\gamma(0)$ then $\gamma|_{[0,t]}$ is not minimal for $t > t_0$. To see this let Y be a non zero Jacobi field along γ with $Y(0) = 0$, $Y(t_0) = 0$. Extend Y to be zero on $[t_0, t]$, then clearly $I(Y,Y) = 0$ on $[0,t]$. Choose $\sigma > 0$ so that $\gamma|_{[t_0 - \sigma, t_0 + \sigma]}$ has no conjugate points. Define X on $[0,t]$ by $X|_{[0, t_0 - \sigma]} = Y$, $X|_{[t_0 - \sigma, t_0 + \sigma]}$ is the unique Jacobi field with $X(t_0 - \sigma) = Y(t_0 - \sigma)$, $X(t_0 + \sigma) = 0$, and $X|_{[t_0 + \sigma, t]} = 0$. By the index lemma 2.25 we have $I(X,X) < I(Y,Y)$ on $[t_0 - \sigma, t_0 + \sigma]$ and $X = Y$ outside this interval i.e. $I(X,X) < 0$ on $[0,t]$. By the second variation formula (2.21) X defines a variation in which γ is the longest curve. Similarly if (ii) holds γ is not minimal on $[0,t]$, $t > t_0$ (cf. (2.8) and discussion).

Let $S(M) = \{v \in TM \mid \|v\| = 1\}$, be the unit tangent bundle of M. It is then a fairly straightforward consequence of (3.14) that the map

$$S(M) \to R \cup \{\infty\} , \quad v \to t_v$$

is continuous. Here $t_v = \infty$ if $\gamma_v(t) = \exp(tv)$ has no cut points.

The <u>injectivity radius</u> at p is defined as

$$\text{inj}_p = d(p, C(p))$$

and the <u>injectivity radius of</u> M as

$$\text{inj}(M) = \inf\{\text{inj}_p \mid p \in M\} \ .$$

Clearly $\text{inj}(M)$ is the largest radius of balls on which $\exp$ is an imbedding. Moreover, if

$$U_p = \{t \cdot v \in T_p M \mid t < t_v\}$$

then $\exp \mid U_p$ is an imbedding onto $M \setminus C(p)$, and $C(p)$ has measure zero. Thus for any measurable set $A \subset M$

$$\text{vol}(A) = \text{vol}(A \setminus C(p))$$

(3.15)
$$= \int_{U_p \cap \exp_p^{-1}(A)} \det(\exp_{p*}) \, d\lambda \ .$$

Now let $V_r(p) = \text{vol}(B_r^M(p))$ be the volume of the open metric ball in M with radius r and center p. Similarly $V_r^\kappa = \text{vol}(B_r^{M_\kappa^n}(p))$, where M_κ^n is the simply connected space form with curvature κ.

<u>3.16 Bishop-Gromov.</u> Let M be a complete n-dimensional riemannian manifold with $\text{Ric}(M) \geq (n-1)\kappa$.

(i) For any $p \in M$, $\bar{p} \in M_\kappa^n$ and unit vectors $v \in T_p M$, $\bar{v} \in T_{\bar{p}} M_\kappa^n$ the function

$$t \to f(t) = \frac{\det(\exp_{p*tv})}{\det(\exp_{\bar{p}*t\bar{v}})}$$

is monotonically non-increasing in t as long as t_v is a regular point for $\exp_p$.

(ii) The inequality

$$\frac{V_r(p)}{V_r^\kappa} \geq \frac{V_R(p)}{V_R^\kappa}$$

holds for all $R \geq r > 0$.

Proof. (i) Let $\iota : T_{\bar p} M_\kappa \to T_p M$ be an isometry such that $\iota(\bar v) = v$ and consider the map $F = \exp_p \circ \iota \circ \exp_{\bar p}^{-1} : B_{\pi/\sqrt{\kappa}}(0) \to M$. Then $f(t) = \det(F_{*\exp(t\bar v)})$. Now suppose $\bar v_1, \ldots, \bar v_{n-1}$ span $\bar v^{\perp}$ and $v_i = \iota(\bar v_i)$. Let $Y_i, \bar Y_i$ denote the corresponding Jacobi fields along $\gamma = \gamma_v$, $\bar\gamma = \gamma_{\bar v}$ i.e. $Y_i(0) = 0$, $Y_i'(0) = v_i$ etc. Then as long as γ contains no conjugate points (cf. 3.14) we have

$$f(t) = \frac{|Y_1(t) \ldots Y_{n-1}(t) \dot\gamma(t)|}{|\bar Y_1(t) \ldots \bar Y_{n-1}(t) \dot{\bar\gamma}(t)|} = \frac{|Y_1(t) \ldots Y_{n-1}(t)|}{|\bar Y_1(t) \ldots \bar Y_{n-1}(t)|}.$$

Fix r and choose $v_i = \iota(\bar v_i)$ so that $Y_1(r), \ldots, Y_{n-1}(r)$ form an orthonormal basis for $\dot\gamma(r)^\perp$. Then

$$|Y_1 \ldots Y_{n-1}|'(r) = \sum_{i=1}^{n-1} <Y_i, Y_i'>(r)$$

$$= \sum_{i=1}^{n-1} I(Y_i, Y_i),$$

where I is the index form on $\gamma : [0,r] \to M$.
By the index lemma (2.25)

$$|Y_1 \ldots Y_{n-1}|'(r) \le \sum_{i=1}^{n-1} I(X_i, X_i)$$

for any vector fields X_i along γ with $X_i(0) = 0$ and $X_i(r) = Y_i(r)$. In particular for $X_i = g \cdot E_i$ where E_i is parallel along γ and $E_i(r) = Y_i(r)$ this yields

$$|Y_1 \ldots Y_{n-1}|'(r) \le \sum_{i=1}^{n-1} \int_0^r (g'(t)^2 - K(\sigma_i(t))g(t)^2) dt$$

where $\sigma_i(t) = \text{span}(E_i(t), \dot\gamma(t))$. By our assumption $\text{Ric}(M) \ge (n-1)\kappa$

$$|Y_1 \ldots Y_{n-1}|'(r) \le (n-1) \int_0^r ((g')^2 - \kappa g^2) dt$$

for any g with $g(0) = 0$ and $g(r) = 1$. However, $\int_0^r ((g')^2 - \kappa g^2) dt = I(g\bar E, g\bar E)$ for any parallel field $\bar E$ along $\bar\gamma : [0,r] \to M_\kappa$. Again by (2.25) the optimal choice for g is when $g\bar E = \bar Y$ is a Jacobi field in M_κ along $\bar\gamma$ i.e.

$$|Y_1 \ldots Y_{n-1}|'(r) \le (n-1) \, I(\bar{Y},\bar{Y})$$

and thus

(*) $\quad f'(r) \le \dfrac{|\bar{Y}_1 \ldots \bar{Y}_{n-1}|(r) \cdot (n-1) \, I(\bar{Y},\bar{Y}) - |\bar{Y}_1 \ldots \bar{Y}_{n-1}|'(r)}{|\bar{Y}_1 \ldots \bar{Y}_{n-1}|^2(r)}$.

Now let

$$g_\kappa(t) = \begin{cases} \dfrac{\sin(\sqrt{\kappa}\, t)}{\sqrt{\kappa}}, & \kappa > 0 \\ t, & \kappa = 0 \\ \dfrac{\sinh(\sqrt{-\kappa}\, t)}{\sqrt{-\kappa}}, & \kappa < 0 \end{cases}$$

and choose parallel fields $\bar{E}_i$ s.t. $\bar{Y}_i = g_\kappa \bar{E}_i$.
Then clearly

$$|\bar{Y}_1 \ldots \bar{Y}_{n-1}| = g_\kappa^{n-1} \cdot |\bar{E}_1 \ldots \bar{E}_{n-1}|$$

and hence

$$\dfrac{|\bar{Y}_1 \ldots \bar{Y}_{n-1}|'}{|\bar{Y}_1 \ldots \bar{Y}_{n-1}|} = (n-1) \, \dfrac{g_\kappa'}{g_\kappa} .$$

Since $\bar{Y} = \dfrac{g_\kappa}{g_\kappa(r)} \bar{E}$ we have

$$I(\bar{Y},\bar{Y}) = \langle \bar{Y}, \bar{Y}' \rangle (r) = \dfrac{g_\kappa'(r)}{g_\kappa(r)}$$

and thus the right hand side of (*) is zero. This completes the proof of (i).

(ii) Note that the numerator is $f(t)$ is independent of $\bar{v}$. Let $d\sigma_r$ denote the standard measure on the sphere $S_r(0) \subset \mathbb{R}^n$. Then by (i)

$$r \to \bar{h}(r) = \dfrac{h(r)}{h_\kappa(r)} = \dfrac{\int_{S_r(0) \cap U_p} \det(\exp_{p *} r v) \, d\sigma_r}{\int_{S_r(0)} \det(\exp_{\bar{p} *} r\bar{v}) \, d\sigma_r}$$

is a non-increasing function in r; the possible existence of cut points only helps matters. From this it follows that also

$$r \to \frac{\int_0^r h(t)dt}{\int_0^r h_\kappa(t)dt}$$

is a non-increasing function in r. To see this let $R \geq r > 0$. Then

$$\int_0^r h \int_r^R h_\kappa = \int_0^r \bar{h} h_\kappa \int_r^R h_\kappa$$

$$\geq \bar{h}(r) \int_0^r h_\kappa \int_r^R h_\kappa$$

$$= \int_0^r h_\kappa \int_r^R \bar{h}(r) h_\kappa$$

$$\geq \int_0^r h_\kappa \int_r^R \bar{h} h_\kappa$$

$$= \int_0^r h_\kappa \int_r^R h$$

or

$$\int_0^r h \int_0^R h_\kappa = \int_0^r h \int_0^r h_\kappa + \int_0^r h \int_r^R h_\kappa$$

$$\geq \int_0^r h_\kappa \int_0^r h + \int_0^r h_\kappa \int_r^R h$$

$$= \int_0^r h_\kappa \int_0^R h \ .$$

Since $V_r(p) = \int_0^r h$, $V_r^\kappa = \int_0^r h_\kappa$ (cf. 3.14) the statement in (ii) follows.

Note that both (3.5) and (3.16 ii) are global comparison theorems. This makes them very powerful as we will see in the next section.

4. GEOMETRY AND TOPOLOGY.

The study of relations between geometry and topology of riemannian manifolds is of fundamental importance in global riemannian geometry. We want to give an impression of this topic by presenting a selection of central results. All of them relate to manifolds with lower curvature bounds, in particular to manifolds with non-negative or positive curvature.

There are subtle topological obstructions for manifolds, M, to carry metrics of positive scalar curvature. One such obstruction is that the fundamental group $\pi_1(M)$ cannot be "too large" in some sense. The methods used to deal with these problems are not based on comparison theory (as in § 3) and are generally more analytic in nature. Here we only refer to the beautiful work of Gromov-Lawson [GL], Rosenberg [R], Hitchin [Hi], Kazdan-Warner [KW], and Schoen-Yau [SY_1] (cf. the lectures by J. Kazdan).

For the Ricci curvature it is natural to begin with the classical

4.1 Bonnet-Myers. Let M^n be a complete riemannian manifold with $Ric(M) \geq (n-1) \cdot 1$. Then M is compact, in fact $diam(M) \leq \pi$.

<u>Proof.</u> Let $\gamma : [0,\ell] \to M$ be a normal geodesic with $\ell > \pi$. Choose an orthonormal basis of parallel fields $E_1, \ldots, E_{n-1}$ along γ perpendicular to $\dot\gamma$. Set $X_i = \sin\left(\frac{\pi}{\ell} t\right) E_i$, then (cf. 2.22).

$$I(X_i, X_i) = \int_0^\ell \left(\frac{\pi}{\ell}\right)^2 \cos^2\left(\frac{\pi}{\ell} t\right) - \sin^2\left(\frac{\pi}{\ell} t\right) K(\sigma_i(t)) ,$$

where $\sigma_i(t) = span(X_i(t), \dot\gamma(t))$. By assumption,

$$\sum_{i=1}^{n-1} I(X_i, X_i) \leq (n-1) \int_0^\ell \left(\frac{\pi}{\ell}\right)^2 \cos^2\left(\frac{\pi}{\ell} t\right) - 1 \cdot \sin^2\left(\frac{\pi}{\ell}\right) t$$

$$< (n-1) \left(\frac{\pi}{\ell}\right)^2 \int_0^\ell \cos^2\left(\frac{\pi}{\ell} t\right) - \sin^2\left(\frac{\pi}{\ell} t\right) = 0 .$$

In particular there is an i such that

$$I(X_i, X_i) < 0 ,$$

and hence γ is not minimal by (2.21).

Let M be as in (4.1). By applying (4.1) to the universal covering $\tilde{M}$ of M one gets immediately

(4.2) $\pi_1(M)$ is finite.

Here $\pi_1(M)$ denotes the fundamental group of M.

The next result is a typical result of what is sometimes referred to as a **rigidity theorem**. Our proof is due to Shiohama [S] (cf. also Itokawa [I]).

4.3 Cheng [Cg]. Let M^n be a complete riemannian manifold with $\text{Ric}(M) \geq (n-1)$. If the diameter of M satisfies $\text{diam}(M) = \pi$, then M is isometric to $S^n(1)$.

Proof. Let $p, \bar{p} \in M$ be at maximal distance, i.e. $d(p,\bar{p}) = \text{diam}(M) = \pi$. Then $B_{\pi/2}(p) \cap B_{\pi/2}(\bar{p}) = \emptyset$ and thus

$$\text{vol}(M) \geq V_{\pi/2}(p) + V_{\pi/2}(\bar{p}) .$$

On the other hand (3.16 ii) implies that

$$\frac{V_{\pi/2}(p)}{V^1_{\pi/2}} \geq \frac{V_\pi(p)}{V^1_\pi} = \frac{\text{vol}(M)}{\text{vol}(S^n(1))}$$

i.e. $V_{\pi/2}(p) \geq \frac{1}{2} \text{vol}(M)$ and similarly $V_{\pi/2}(\bar{p}) \geq \frac{1}{2}\text{vol}(M)$. In other words, if $f_p(r) = V_r(p)/V^1_r$, then $f_p(\pi/2) = f_p(\pi) = f_{\bar{p}}(\pi) = f_{\bar{p}}(\pi/2)$. From the monotonicity property of f_p it follows $f_p(r) = f_p(\pi)$, i.e. $V_r(p) = f_p(\pi) \cdot V^1_r$ for $r \in [\pi/2, \pi]$. Moreover, $B_r(p) \cap B_{\pi-r}(\bar{p}) = \emptyset$ and hence

$$\text{Vol}(M) \geq V_r(p) + V_{\pi-r}(\bar{p})$$

$$= f_p(r) V^1_r + f_{\bar{p}}(\pi-r) V^1_{\pi-r}$$

$$\geq f_p(\pi) V^1_r + f_{\bar{p}}\left(\frac{\pi}{2}\right) V^1_{\pi-r} , \quad r \in [\pi/2, \pi]$$

$$= f_p(\pi) (V^1_r + V^1_{\pi-r})$$

$$= \text{Vol}(M) .$$

It follows that $f_{\bar{p}} = f_p = 1$ on $[0, \pi]$. By (3.15) and (3.16) $U_p = B_\pi(0) \subset T_p M$ and $U_{\bar{p}} = B_\pi(0) \subset T_{\bar{p}} M$. In particular any normal geodesic $\gamma : [0, \pi] \to M$ with $\gamma(0) = p$ is minimal. Therefore, $I(X,X) \geq 0$ for any vector field X along γ with $X(0) = 0$, $X(\pi) = 0$ and $X \perp \dot{\gamma}$. Thus for any parallel field E along γ with $\|E\| = 1$ and $E \perp \dot{\gamma}$ we have that

$Y = \sin \cdot E$ is a Jacobi fiels along γ (cf. proof of (4.1)). This has two immediate consequences. One is that $\exp_{p*}$ annihilates all tangent vectors to $C_p = S_\pi(0) \subset T_p M$ and hence $C(p) = \bar{p}$. Another is that $K(\sigma) = 1$ for any plane containing $\dot{\gamma}$. It is now obvious that if $x \in S^n(1)$ and $\iota : T_x S^n(1) \to T_p M$ is an isometry then $F : S^n(1) \to M$ defined by

$$F(y) = \begin{cases} \exp_p \circ \iota \circ \exp_x^{-1}(y) &, \quad y \neq -x \\ \bar{p} &, \quad y = -x \end{cases}$$

is an isometry.

It is curious that for simply connected manifolds there are no general topological obstructions known for carrying metrics with positive Ricci curvature other than those for positive scalar curvature. For properties on the fundamental group of a manifold with non-negative Ricci curvature we refer to Cheeger-Gromoll [CG$_1$], Milnor [M$_2$], Gromov [GLP].

In dimension 3 we mention two outstanding results

<u>4.4 Schoen-Yau</u> [SY$_2$] Let M^3 be a complete non-compact riemannian manifold with $\mathrm{Ric}(M) > 0$. Then M is diffemorphic to R^3.

<u>4.5 Hamilton</u> [Ha]. Let M^3 be a compact riemannian manifold with $\mathrm{Ric}(M) > 0$. Then M is diffeomorphic to a 3-dimensional space form $S^3(1)/\Gamma$.

Neither is based on comparison theory as in § 3, although in the proof of 4.4 (cf. also [SY$_1$]) the geometry of minimal surfaces plays a role like geodesics. In (4.5) it is actually proved that the metric can be deformed through metrics of positive Ricci curvature to one of constant curvature. The proof is based on heat equation methods.

In the rest of this section we confine ourselves to results on the sectional curvature.

The following notion is useful in general:

For any closed set $A \subset M$ we say that $x \in M \setminus A$ is a <u>critical point</u> for $d_A = d(A, \cdot)$ if for any $V \in T_x M$ there is a shortest geodesic γ from x to A such that $\angle(\dot{\gamma}(0), V) \leq \pi/2$.

A non-critical point is called <u>regular</u>.

Observe that if $x \in M \setminus A$ is regular then all nearby points are regular as well. In fact. there is a unit vector field V in a neighborhood U of x such that

$$\measuredangle\,(\dot{\gamma}(0),\ V) \geq \frac{\pi}{2} + \varepsilon$$

for some $\varepsilon > 0$ and all shortest geodesics γ from points in U to A. It follows e.g. by (2.8) that d_A is strictly increasing along the integral curves of V. In particular

(4.6) A local maximum point $x \in M$ for d_A is a critical point.

Now let us consider first manifolds M with $K \geq 1$. By (4.1) and (4.3) any such M is compact, in fact $\text{diam}(M) \leq \pi$, where equality holds if M is isometric to $S^n(1)$. The last statement was much prior to (4.3) by Toponogov as an application of (3.5).

As a generalization of the classical sphere theorem of Rauch-Berger-Klingenberg (cf. [CE]) one has [GS]:

<u>4.7 Diameter Sphere Theorem.</u> Let M^n be a complete riemannian manifold with $K \geq 1$ and $\text{diam}(M) > \pi/2$. Then M is homeomorphic to S^n.

<u>Proof.</u> Choose $p, \bar{p} \in M$ at maximal distance, i.e. $d(p,\bar{p}) = \text{diam}(M) > \pi/2$. Then $\bar{p}$ is uniquely determined by p. Otherwise suppose $d(p,\bar{p}') = \text{diam}(M)$ and let γ_1 be a minimal geodesic from $\bar{p}$ to $\bar{p}'$. Since $\bar{p}, \bar{p}'$ are critical for d_p by (4.6) we can choose minimal geodesics γ_2 from $\bar{p}'$ to p and γ_3 from p to $\bar{p}$ such that $\alpha_i \leq \pi/2$, $i=2,3$. This is impossible by (3.5 A).

Now let $(\gamma_1, \gamma_2, \alpha)$ be a hinge at $x \in M \setminus \{p, \bar{p}\}$ where γ_1 is a minimal geodesic from $\bar{p}$ to x and γ_2 is a minimal geodesic from x to p. Since $d(p,\bar{p}) = \text{diam}(M) > \pi/2$ it follows from (3.5 B) that $\alpha > \pi/2$. This implies that no $x \in M \setminus \{p,\bar{p}\}$ is a critical point for d_p. By partition of unity we can construct a unit vector field V on $M \setminus (B_\varepsilon(p) \cup B_\varepsilon(\bar{p}))$ such that V is perpendicular to the boundary components and such that d_p is strictly increasing along the integral curves of V. Thus M is homeomorphic to S^n.

For manifolds M with $K \geq 1$ and $\text{diam}(M) = \pi/2$ there is a rigidity theorem [GG] which extends the classical rigidity theorem of Berger.

Note that the "<u>diameter-pinching</u>"

(4.8) $$\partial_M = \min K \cdot \frac{\operatorname{diam}(M)^2}{\pi^2}$$

is invariant under scalings of the riemannian metric. By (4.1) $\partial_M \leq 1$ where by (4.3) equality holds if M is a round sphere. If $\partial_M > 1/4$ then M is homeomorphic to S^n by (4.7).

A fundamental result of Gromov [G₁] (cf. also Abresch [A]) can be stated as

4.9 Gromov. For any number $D \leq 1$ and integer $n \geq 2$ there is a number $C(n,D)$ such that: If M is a compact n-dimensional manifold with $\partial_M \geq D$, then

$$\sum_{i=0}^{n} \beta_i(M) \leq C(n,D) ,$$

where $\beta_i(M)$ is the i'th Betti number of M (any field as cofficients). The idea of critical points for distance functions plays an essential role at the various steps of the proof. The geometric part of the proof is based upon (3.5) and (3.16). The topological part requires prerequisites, which we have not elaborated on here.

There is an analogous but much simpler result, also due to Gromov [G₂], for the number of generators of the fundamental group, $\pi_1(M)$:

4.10 Gromov. For D,n as in (4.9) there is a number $C'(n,D)$ such that: If M^n satisfies $\partial_M \geq D$ then $\pi_1(M)$ can be generated by $\leq C'(n,D)$ elements.

In both cases notice that diam(M) does not enter, when $K \geq 0$. The proof of (4.10) is a beautifully simple application of (3.5). For simplicity, we present only:

Proof of (4.10) for $K \geq 0$. Assume M compact (cf. (4.11)) and let $\tilde{M}$ be the universal covering of M. Consider elements $g \in \pi_1(M)$ as deck transformations $g: \tilde{M} \to \tilde{M}$. Fix $\tilde{p} \in \tilde{M}$ and choose a set of generators $g_1, \ldots, g_k \in \pi_1(M)$ such that

(*) $$\sum_{i=1}^{k} d(\tilde{p}, g_i(\tilde{p}))$$

is minimal. If γ_i are minimal geodesics in $\tilde{M}$ from $\tilde{p}$ to $g_i(\tilde{p})$ then

$\angle(\dot{\gamma}_i(0), \dot{\gamma}_j(0)) \geq \pi/3$ for $i \neq j$. To see this suppose $\angle(\dot{\gamma}_i(0), \dot{\gamma}_j(0)) < \pi/3$ for some $i \neq j$. Then by (3.5 B) $d(g_i(\tilde{p}), g_j(\tilde{p})) < \max\{d(\tilde{p}, g_i(\tilde{p})), d(\tilde{p}, g_j(\tilde{p}))\}$, which contradicts the minimality of (*) (replace g_j by $g_i^{-1} g_j$). However, if $v_1, \ldots, v_k \in S^{n-1}(1)$ satisfy $d(v_i, v_j) \geq \pi/3$, then $B_{\pi/6}(v_i) \cap B_{\pi/6}(v_i) \cap B_{\pi/6}(v_j) = \emptyset$, and hence $k \cdot V_{\pi/6}^1 \leq \text{vol}(S^{n-1})$. This gives a bound for k.

The remaining part of this section is devoted to a presentation of the so-called <u>Soul-theorem</u> by Cheeger-Gromoll-Meyer (cf. also [Po]):

4.11 Cheeger, Gromoll, Meyer. Let M^n be a complete non-compact riemannian manifold with $K \geq 0$. Then there is a compact totally geodesic submanifold $S \subset M$ such that M is diffeomorphic to $TS^{\perp}$. If $K > 0$ then M is diffeomorphic to R^n.

There are several steps in the proof of (4.11).

First recall that a <u>ray</u> $\gamma: [0, \infty) \to M$ is a normal geodesic minimal on all finite segments. For each $p \in M$ there is a ray γ with $\gamma(0) = p$: Simply choose a sequence $p_i \in M$ with $d(p, p_i) \to \infty$ and let γ_i be a minimal geodesic from p to p_i. If v is an accumulation point of the unit vectors $\dot{\gamma}_i(0)$, then clearly $\gamma = \gamma_v$ is a ray.

Given a ray $\gamma: [0, \infty) \to M$ we define

(4.12)
$$B_\gamma = \bigcup_t B_t(\gamma(t)),$$
$$H_\gamma = M \setminus B_\gamma.$$

H_γ is called the <u>half-space</u> of γ at $\gamma(0) \in H_\gamma$.

Recall that $C \subset M$ is called <u>totally convex</u> if whenever $p, q \in C$ and γ is any geodesic from p to q, then $\gamma \subset C$.

<u>Step 1.</u> For any ray γ, H_γ is totally convex.

<u>Proof.</u> Suppose H_γ is not totally convex and let $\gamma_0: [0,1] \to M$ be a geodesic with $\gamma_0(0), \gamma_0(1) \in H_\gamma$, but $\gamma_0(s) \in B_\gamma$ for some $s \in (0,1)$. Then $q = \gamma_0(s) \in B_{t_0}(\gamma(t_0))$ for some $t_0 > 0$ and hence $q \in B_t(\gamma(t))$ for $t > t_0$. In fact, setting

$$t_0 - \varepsilon = d(q, \gamma(t_0)), \quad \varepsilon > 0$$

we have

$$d(q,\gamma(t)) \leq d(q,\gamma(t_0)) + d(\gamma(t_0),\gamma(t))$$
$$= t_0 - \varepsilon + t - t_0$$
$$= t - \varepsilon$$

for all $t \geq t_0$.

For each $t \geq t_0$ let $\gamma_0(s_t)$ be a closest point to $\gamma(t)$ on γ_0, and let γ_1^t be a minimal geodesic from $\gamma(t)$ to $\gamma_0(s_t)$. From the above inequality

$$L(\gamma_1^t) \leq t - \varepsilon$$

for all $t \geq t_0$. On the other hand, consider the hinge $(\gamma_1^t, \gamma_2^t, \pi/2)$, where $\gamma_2^t = \gamma_0^{-1}|[s_t, 0]$. By (3.5 B)

$$d(\gamma(t),\gamma_0(0)) \leq (L(\gamma_2^t)^2 + L(\gamma_1^t)^2)^{\frac{1}{2}}$$

and hence

$$t^2 \leq d(\gamma(t),\gamma_0(0))^2 \leq L(\gamma_0)^2 + (t - \varepsilon)^2$$

for all $t \geq t_0$. This is impossible for t large.

Now fix $p \in M$ and consider all rays γ from p. For each $t \geq 0$ and ray γ let γ_t be the ray $s \to \gamma(t+s)$. Define

$$C_t = \bigcap_\gamma H_{\gamma_t}.$$

<u>Step 2.</u> C_t is a compact totally convex set for all $t \geq 0$. Moreover,

(i) $$C_{t_2} \supset C_{t_1} \quad \text{for } t_2 \geq t_1, \quad \text{in fact}$$

$$C_{t_1} = \{q \in C_{t_2} | d(q, \partial C_{t_2}) \geq t_2 - t_1\}$$

and in particular

$$\partial C_{t_1} = \{q \in C_{t_2} | d(q, C_{t_2}) = t_2 - t_1\}.$$

(ii) $$\bigcup_{t \geq 0} C_t = M.$$

(iii) $$p \in \partial C_0.$$

Proof. Clearly, C_t is a closed totally convex set with $p \in C_t$ for all $t \geq 0$. Then, if C_t were not compact we could find a ray γ at p entirely contained in C_t. By definition of C_t this is impossible. The remaining part of the proof is essentially an exercise in the triangle inequality (cf. [CE]).

Note that $\text{int } C_t \neq \emptyset$ for $t > 0$. This is not necessarily true for C_0, and we need to understand the structure of convex sets in general.

Recall that there is a continuous function $r : M \to R_+$, the <u>convexity radius</u>, such that $B_{r(p)}(p)$ is <u>strongly convex</u> for all $p \in M$, i.e. for any $q, q' \in B_{r(p)}(p)$ there is a unique minimal geodesic $\gamma_{qq'}$ from q to q' and $\gamma_{qq'} \subset B_{r(p)}(p)$. In general, we say that $C \subset M$ is <u>convex</u> if for any $p \in \bar{C}$ there is $0 < \varepsilon(p) < r(p)$ such that $C \cap B_{\varepsilon(p)}(p)$ is strongly convex. A totally convex set is of course connected and convex in this sense. Also the closure of a convex set is convex.

Assume C is convex. Let $0 \leq k \leq n$ be the largest integer such that the collection $\{N_\alpha\}$ of smoothly imbedded k-dimensional submanifolds of M with $N_\alpha \subset C$ is non-empty. Let $N = \bigcup_\alpha N_\alpha$.

<u>Claim.</u> N is a smooth connected totally geodesic submanifold of M and $C \subset \bar{N}$.

To see this let $p \in N$. Then $p \in N_\alpha$ for some α and we may choose a neighborhood $U \subset N_\alpha \cap B_{\frac{1}{2}\varepsilon(p)}(p)$ of p in N_α and $0 < \delta < \frac{1}{2}\varepsilon(p)$ such that the normal exponential map $\exp : TU^\perp \to M$ restricted to vectors of length $< \delta$ is a diffeomorphism onto a neighborhood T_δ of p in M. By convexity of C and the choice of k it follows that $C \cap T_\delta \setminus U = \emptyset$. In particular $N \cap T_\delta = U$ and N is a k-dimensional totally geodesic submanifold. A slightly more detailed analysis along the same lines shows that N is connected and $C \subset \bar{N}$ (cf. [CE], [CG$_2$]). Moreover, $\bar{N}$ is a topological manifold with possibly empty boundary $\partial \bar{N} = \bar{N} \setminus N$.

The <u>tangent cone</u> at $p \in C$ is by definition the set

$$(*) \quad \{V \in T_p M \mid \exp\left(t \frac{V}{\|V\|}\right) \in N \text{ for some } 0 < t < \varepsilon(p)\} \cup \{0\} \subset T_p M.$$

If $p \in N$ this is just $T_p N = T_p C$. Otherwise it is a cone in the k-dimensional subspace generated by it. Suppose C is closed, $p \in \partial C$ and $q \in \text{int } C = N$ such that $d(q,p) = d(q, \partial C)$. The tangent cone at p minus 0 is then the open half-space

(**) $\{V \in T_pM \mid \angle(V,\dot{\gamma}(0)) < \pi/2\}$

where γ is a minimal geodesic from p to q.

For complete proofs of the above statements we refer to [CE], [CG$_2$].

For a closed convex set C with $\partial C \neq \emptyset$, we set

$$C^a = \{p \in C \mid d(p,\partial C) \geq a\},$$

$$C^{max} = \bigcap_{C^a \neq \emptyset} C^a.$$

Step 3. Let $C \subset M$ be a closed totally convex set with $\partial C \neq \emptyset$. Then

(i) C^a is closed and totally convex, $a \geq 0$

(ii) $\dim C^{max} < \dim C$.

Moreover, if $K > 0$, then

(iii) C^{max} is a point.

This is a direct consequence of the following.

Claim. The function $\psi = d(\cdot, \partial C): C \to \mathbb{R}$ is <u>concave</u>, i.e. for any normal geodesic $\gamma: [a,b] \to C$

$$\psi(\gamma(st_1 + (1-s)t_2)) \geq s\psi(\gamma(t_1)) + (1-s)\psi(\gamma(t_2))$$

where $t_1, t_2 \in [a,b]$, $s \in [0,1]$. The inequality is strict when $K \geq 0$ is replaced by $K > 0$.

Proof. Let $\gamma: [a,b] \to C$ be a normal geodesic. For $a < s_0 < b$ let $\gamma_{s_0}: [0,d] \to C$ be a minimal geodesic from $\gamma(s_0)$ to ∂C. It suffices to show that on some interval $(s_0 - \delta, s_0 + \delta)$, $\psi \circ \gamma$ is bounded above by the linear function

$$d - \cos \alpha \cdot (s - s_0),$$

where $\alpha = \angle(\dot{\gamma}(s_0), \dot{\gamma}_{s_0}(0))$. For the argument, it is only necessary to consider $s > s_0$, and the three cases $\alpha = \pi/2$, $\alpha > \pi/2$, $\alpha < \pi/2$.

$\underline{\alpha = \pi/2}$: Let E be the parallel field along γ_{s_0} with $E(0) = \dot{\gamma}(s_0)$. By corollary (3.4) of Rauch II there is a $\delta > 0$, such that for $s_0 \leq s \leq s_0 + \delta$ the curve

$$\gamma_s(t) = \exp_{\gamma_{s_0}}(s - s_0) E(t)$$

has length $L(\gamma_s) \leq d$. Since int C is totally geodesic this clearly implies $\psi(\gamma(s)) \leq d = d - \cos \pi/2 \cdot (s - s_0)$.

$\underline{\alpha > \pi/2}$: Choose E so that $E(0)$ is the convex combination of $\dot{\gamma}_{s_0}(0)$ and $\dot{\gamma}(s_0)$ with $E(0) \perp \dot{\gamma}_{s_0}(0)$. As before, $L(\gamma_s) \leq d$ for $s_0 \leq s \leq s_0 + \delta$. Furthermore, $d(\gamma(s), \exp(\cos(\alpha - \pi/2)(s - s_0)) E(0)) \leq -(s - s_0) \cos \alpha$ by (3.5 B). Combining these inequalities yields $\psi(\gamma(s)) \leq d - \cos \alpha \cdot (s - s_0)$.

$\underline{\alpha < \pi/2}$: Let c_s be the minimal geodesic from γ_{s_0} to $\gamma(s)$. If $c_s(0) = \gamma_{s_0}(t_s)$, then $\dot{c}_s(0) \perp \dot{\gamma}_{s_0}(t_s)$. Let E be the parallel field along $\gamma_{s_0}: [t_s, d] \to C$ determined by $E(t_s) = \dot{c}_s(0)$. Applying (3.5 B) to the hinges $(\gamma_{s_0}|[0,t_s], \gamma|[s_0,s], \alpha)$ and $(c_s, \gamma_{s_0}^{-1}|[t_s,0], \alpha/2)$ yields

$$(s - s_0) \cos \alpha \leq t_s.$$

Combining this with the estimates for the curves generated by E gives the desired $\psi(\gamma(s)) \leq d - t_s \leq d - \cos \alpha \cdot (s - s_0)$.

The equality discussion is left to the reader.

Note that by repeated applications of Step 3 one gets from a closed totally convex set C with $\partial C \neq \emptyset$ a totally convex set $S \subset C$ with dim $S <$ dim C and $\partial S = \emptyset$. In particular S is a closed totally geodesic submanifold of M. S is called the <u>soul of</u> C.

<u>Step 4.</u> If S is the soul of C_t from Step 2, then M is diffeomorphic to $TS^\perp$.

<u>Proof.</u> Let $x \in M \setminus S$. By Step 2 either $x \in \partial C_t$ for some $t \geq 0$ or $x \in \text{int } C_0$. By Step 3 either $x \in \partial C_0^a$ for some $a \geq 0$ or $x \in \text{int } C_0^{\max}$. Applying this argument at most finitely many times yields: For any $x \in M \setminus S$

there is a compact totally convex set $C \subset M$ with $\partial C \neq \emptyset$ such that $x \in \partial C$ and $S \subset C$. In particular for any minimal geodesic γ from x to S the vector $\dot{\gamma}(0)$ is in the tangent cone (*) at x. This shows that the function $d_S : M \to R$ has no critical points on $M \setminus S$. As in the proof of (4.7) we can construct a unit vector field V on $M \setminus B_\varepsilon(S)$ such that V is perpendicular to the boundary of the tubular neighborhood $B_\varepsilon(S)$ and d_S is strictly increasing along the integral curves of V. This completes the proof of 4.11.

5. SPACES OF RIEMANNIAN MANIFOLDS.

So far we have considered "individual" riemannian manifolds. Even for this purpose it can be very helpful to view the manifold as a member of a larger collection of riemannian manifolds. A classical example of this is the beautiful Teichmüller theory, in which one studies the <u>moduli space</u>, M_g, of isometry classes of space forms of a surface, T_g, of genus $g \geq 2$ (or equivalently the space of complex structures on T_g). Another similar example is the moduli space of self-dual connections (cf. the lectures by Bourguignon and Taubes).

Here we present some of the foundational work of Gromov [GLP] on metric space structures on collections of riemannian manifolds. In particular we discuss the so-called <u>precompactness</u> and <u>compactness</u> theorem and give applications of the latter.

Already from Teichmüller theory it is clear that one has to allow for singularities on possible limit objects. It is thus natural to begin our discussion in the framework of general metric spaces.

Let X and Y be metric spaces. The <u>dilatation</u> of a map $f : X \to Y$ is the (possibly infinite) number

$$(5.1) \qquad \text{dil } f = \sup\left\{ \frac{d^Y(f(x_1),f(x_2))}{d^X(x_1,x_2)} \ \bigg| \ (x_1,x_2) \in (X \times X)\setminus\Delta \right\}.$$

The <u>local dilatation</u> of f at x is the number

$$(5.2) \qquad \text{dil}_x f = \lim_{\varepsilon \to 0} \text{dil } f\big|_{B_\varepsilon(x)}.$$

If $c = \text{dil } f < \infty$ we say that f is <u>Lipschitz</u> with <u>Lipschitz constant</u> c. A Lipschitz homeomorphism is a homeomorphism f, such that f and f^{-1} are Lipschitz.

The <u>Lipschitz distance</u> between X and Y is the number

$$(5.3) \qquad d_L(X,Y) = \inf\{|\log \text{dil } f| + |\log \text{dil } f^{-1}|\}$$

where the infimum is taken over all Lipschitz homeomorphisms $f : X \to Y$. If no such homeomorphism exists we set $d_L(X,Y) = \infty$.

By equicontinuity clearly

(5.4) If C and Y are compact and $d_L(X,Y) = 0$ then X and Y are isometric.

For subsets A,B of a fixed metric space Z, the classical notion of Hausdorff distance is given as

(5.5) $\quad d_H^Z(A,B) = \inf\{\varepsilon \mid B \subset \bar{B}_\varepsilon(A), A \subset \bar{B}_\varepsilon(B)\}$,

where $\bar{B}_\varepsilon(A) = \{x \in Z \mid d(x,A) \leq \varepsilon\}$. For arbitrary metric spaces X and Y the <u>Hauddorff distance</u> is the number

(5.6) $\quad d_H(X,Y) = \inf\{d_H^Z(f(X), g(Y))\}$,

where the infimum is taken of all metric spaces Z and isometric imbeddings $f : X \to Z$, $g : Y \to Z$.

For compact X and Y endow $Z = X \sqcup Y$ with the obvious metric on X,Y and $d(x,y) = \max\{\text{diam}(X), \text{diam}(Y)\}$ for $x \in X$, $y \in Y$. This shows

(5.7) $\quad d_H(X,Y) < \infty \quad$ for compact X and Y.

5.8 Examples

(i) $\quad d_H([0,1], \mathbb{Q} \cap [0,1]) = 0$

(ii) Let $A = \{a_1, a_2, a_3\}$ with metric $d(a_i, a_j) = 1$ for $i \neq j$ and let $B = \{b\}$ a point. For all isometric imbeddings $f : A \to \mathbb{R}^n$, $g : B \to \mathbb{R}^n$ clearly

$$d_H^{\mathbb{R}^n}(f(A), g(B)) \geq 1/\sqrt{3} .$$

However,

$$d_H(A,B) = 1/2 .$$

We will now relate Hausdorff convergence to Lipschitz convergence of nets. An ε-<u>net</u> of X is a subset $N \subset X$ such that $X \subset B_\varepsilon(N)$. The <u>mesh</u> of N is the number

$$\mu_N = \inf \{d(x,y) \mid (x,y) \in N \times N \setminus \Delta\} .$$

We will use the notation N_X^ε for an ε-net in X.

5.9 (A). Given positive numbers ε, r and μ. For any $\mu' < \mu$, $r' > r$ there exists a $\delta > 0$ such that if

$$d_H(X,Y) < \delta ,$$

then for any r-net N_X^r with mesh $\geq \mu$ there exists an r'-net $N_Y^{r'}$ with mesh $\geq \mu'$ such that

$$d_L(N_X^r, N_Y^{r'}) < \varepsilon.$$

5.9 (B). Given positive numbers ε and D. For any $\varepsilon' < \varepsilon$ there exists a $\delta > 0$ such that if $N_X^{\varepsilon'}, N_Y^{\varepsilon'}$ are ε'-nets with

$$d_L(N_X^{\varepsilon'}, N_Y^{\varepsilon'}) < \delta$$

and $\max\{\text{diam}(X), \text{diam}(Y)\} \leq D$, then

$$d_H(X,Y) \leq \varepsilon.$$

<u>Proof.</u> (A) is a straightforward exercise in applications of the definitions.

In (B) a suitable choice of δ implies the existence of a homeomorphism $F : N_X^{\varepsilon'} \to N_Y^{\varepsilon'}$ with

$$(*) \qquad (1-\eta) d^X(x_n, x_m) \leq d^Y(F(x_n), F(x_m)) \leq (1+\eta) d^X(x_n, x_m)$$

for all $x_n, x_m \in N_X^{\varepsilon'}$ and $\eta \cdot D \leq 2(\varepsilon - \varepsilon')$. Define a distance function d on the disjoint union $Z = X \amalg Y$ which restricts to d^X on X and to d^Y on Y. For $x \in X$, $y \in Y$ we let

$$d(x,y) = \varepsilon - \varepsilon' + \inf\{d^X(x, x_n) + d^Y(y, F(x_n)) \mid x_n \in N_X^{\varepsilon'}\}.$$

To see that $d : Z \times Z \to \mathbb{R}$ is a metric, the only non-trivial inequality is of the type

$$d(x,x') \leq d(x,y) + d(y,x')$$

for $x, x' \in X$ and $y \in Y$. However,

$$d(x,x') \leq d(x, x_n) + d(x_n, x_m) + d(x_m, x')$$

for any $x_n, x_m \in N_X^{\varepsilon'}$, and

$$d(x_n, x_m) \leq d(F(x_n), F(x_m)) + 2(\varepsilon - \varepsilon')$$

by $(*)$. Hence

$$d(x,x') \leq d(x, x_n) + d(F(x_n), y) + d(y, F(x_m)) + d(x_m, x') + 2(\varepsilon - \varepsilon')$$

for all $x_n, x_m \in N_X^{\epsilon'}$. This proves the desired inequality.

Since clearly
$$\bar{B}_\epsilon(X) \supset Y \quad, \quad \bar{B}_\epsilon(Y) \supset X$$
for the obvious imbeddings $X, Y \subset Z$ we get $d_H(X,Y) \leqq \epsilon$.

A diagonal argument based on (5.9 A) shows that for compact X, Y with $d_H(X,Y) = 0$ one has $d_L(X,Y) = 0$. Hence (cf. 5.4)

(5.10) Compact metric spaces X, Y with $d_H(X,Y) = 0$ are isometric.

Thus d_H is a metric on isometry classes of compact metric spaces.

An immediate application of (5.9 B) is the following

(5.11) If X_n are compact metric spaces with $\text{Lim}_L X_n = X$, then $\text{Lim}_H X_n = X$.

The converse of (5.11) is obviously wrong:

5.12 Examples.

(i) Let X be a metric space with distance d. For each $\lambda > 0$ let λX denote the metric space $(X, \lambda d)$. If $\text{diam } X < \infty$ the Hausdorff limit of λX for $\lambda \to 0$ is a point.

(ii) Let $X_n = Y \times Z_n$ and suppose $\text{diam } Z_n \to 0$ as $n \to \infty$. Then $\text{Lim}_H X_n = Y$.

(iii) Let $\pi : E \to M$ be a <u>riemannian submersion</u>, i.e. E and M are riemannian manifolds and π_* restricted to each normal space of the fibers $F = \pi^{-1}(x)$, $x \in M$ is an isometry. Suppose M and F are compact. Let g_t be the riemannian metric on E obtained from the original metric by multiplying the length of vertical vectors by $\sqrt{t}$. Then M is the Hausdorff limit of (E, g_t), $t \to 0$.

(iv) Let
$$X_n = \left\{ (x,y) \in I \times I \mid x = \frac{k}{n} \text{ or } y = \frac{k}{n}, k = 0, \ldots, n \right\}$$
with metric, the infimum of lengths of paths. Then $\text{Lim}_H X_n$ for $n \to \infty$ is the square $I \times I$ with metric induced by the norm
$$\|(x,y)\| = |x| + |y|.$$

The first three examples are examples of collapse. The third is particularly interesting:

If $S^3 \to S^2$ is the Hopf fibration, the curvature K_t of the Berger-Spheres (S^3, g_t) stays bounded!

For non-compact spaces X it is customary to choose a base point $x_0 \in X$ and work with the previous concepts of distances between "pointed spaces" (cf. [GLP]). We will not elaborate on this here.

We now turn our attention to the set of compact riemannian manifolds with the Hausdorff metric.

5.13 Precompactness Theorem. Given $\kappa \in \mathbb{R}$ and $D > 0$. The set of all compact riemannian n-manifolds, M^n, with $\mathrm{Ric}(M) \geq (n-1)\kappa$ and $\mathrm{diam}(M) \leq D$ is precompact in the Hausdorff metric.

<u>Proof.</u> Given $\varepsilon > 0$ we need to find finitely many manifolds $M_1, \ldots, M_k$ with Ricci-curvature $\geq (n-1)\kappa$ and diameter $\leq D$ so that any such manifold M satisfies

$$d_H(M, M_\ell) < \varepsilon$$

for some $\ell \in \{1, \ldots, k\}$.

For any M with $\mathrm{Ric}(M) \geq (n-1)\kappa$ and $\mathrm{diam}(M) \leq D$ let $N(r, M)$ be the maximal number of disjoint r-balls in M. By (3.16 ii)

$$N(r, M) \leq \frac{V_D^\kappa}{V_r^\kappa}.$$

Fix r and $n \in \{1, \ldots, V_D^\kappa / V_r^\kappa\}$ and consider the collection of M's with $N(r, M) = n$. The centers $\{x_1, \ldots, x_n\}$ of disjoint r-balls in M clearly form a 2r-net in M. The bijection $\{1, \ldots, n\} \to \{x_1, \ldots, x_n\} \subset M$ induces a distance d_M on $\{1, \ldots, n\}$. If $A = \{(i,j) \mid 1 \leq i < j \leq n\}$ then d_M is determined by $d_M : A \to [2r, D]$. Now the space of functions $[2r, D]^A$ is compact in the product topology. Thus the set of functions $d_M \in [2r, D]^A$ is precompact, i.e. given $\varepsilon' > 0$ we cand find $M_1, \ldots, M_k$ such that for any M there is an ℓ with

$$|d_M(i,j) - d_{M_\ell}(i,j)| < \varepsilon'$$

for all $(i,j) \in A$. In particular

$$d_L\left(\{x_1, \ldots, x_n\}, \{x_1^\ell, \ldots, x_n^\ell\}\right) < 2 \log\left(1 + \frac{\varepsilon'}{r}\right).$$

Now for given ε and $2r < \varepsilon$ choose ε' so that $2\log(1+\varepsilon'/r) < \delta$ from (5.9 B). Then

$$d_H(M,M_\ell) < \varepsilon$$

follows from the above and (5.9 B).

Note that by Bonnet-Myers' theorem (4.1) the set of all complete riemannian manifolds M satisfying Ric(M) $\geq$ (n-1) is precompact in the Hausdorff metric. This suggests a finiteness theorem of some kind .... which one is not at all clear! Compare this with (4.9).

In view of (5.13) it is important to understand limit objects X of Hausdorff converging sequences of complete riemannian manifolds. Not much can be said in general except that a complete X is a so-called <u>length space</u>, i.e. the distance between points of X is the infimum of lengths of curves joining the points (cf. [GLP]).

Consider now the set of compact riemannian n-manifolds M with bounded curvature $|K_M| \leq 1$, diam(M) $\leq$ D and injectivity radius bounded from below inj(M) $\geq \iota$. By (5.13) this set is of course precompact in the Hausdorff metric. Moreover, the closure of this set consists of n-manifolds with some kind of "weak" riemannian metric.

Note that by (3.16) inj(M) $\geq \iota$ implies that there is a lower bound for the volume, Vol(M) $\geq$ V. On the other hand, it is a theorem of Cheeger (cf. [CE] and [KH]) that $|K_M| \leq 1$, diam(M) $\leq$ D and Vol(M) $\geq$ V imply a lower bound inj(M) $\geq \iota$.

In this set of manifolds it follows by comparison theory that if two manifolds are Hausdorff close they must be Lipschitz close. A Lipschitz diffeomorphism is constructed from Lipschitz close nets with small mesh. A useful technical tool for such a construction is the general non-linear notion of "<u>center of mass</u>" introduced in [GK] (cf. also [BK]). Together with (5.13) this yields a proof of Cheeger's finiteness theorem (cf.[Pe$_1$]).

5.14 Cheeger. Given positive numbers D and V. There is only a finite number of diffeomorphism classes of compact riemannian manifolds M such that $|K_M| \leq 1$, diam(M) $\leq$ D and Vol(M) $\geq$ V.

A proof of Gromov's compactness theorem along these lines became clear in discussions at the Geometry-meeting in Luminy, July 1984. The essential analytical tool is to replace exponential coordinates in the center of mass construction by so-called harmonic coordinates (cf. [DK], [JK]). The proof is nontrivial and will not be given here. We refer to the detailed expositions in [GW] and [Pe$_2$].

5.15 Compactness Theorem. Given a sequence $\{M_\ell\}$ of compact riemannian n-manifolds with $|K| \leq 1$, diam $\leq D$ and vol $\geq V$. Then $\{M_\ell\}$ has a subsequence $\{M_k\}$ which converges in the Lipschitz metric to an n-manifold M with "riemannian" metric, g_∞, of class $C^{1,\alpha}$ for every $\alpha \in (0,1)$. Here a $C^{1,\alpha}$ metric is by definition a metric for which the local coefficients g_{ij} are of class $C^{1,\alpha}$.

In (5.15) we can view g_∞ as a limit of riemannian metrics g_k on M where (M, g_k) is isometric to M_k. With this interpretation we have for all $p \in M$:

(i) $d_k(p, \cdot)$ converges uniformly to $d_\infty(p, \cdot)$.

(ii) $\exp_p^k$ converges uniformly to $\exp_p^\infty$ on compact sets.

(iii) $\exp_p^\infty$ is Lipschits. In particular $d\exp_p^\infty$ exists almost everywhere, and $d\exp_p^k$ converges to $d\exp_p^\infty$ almost everywhere.

Remark. This result is optimal in the sense that it fails for $\alpha = 1$, [P$_2$]. Nevertheless there is a notion of curvature almost everywhere.

Warning: Since g_∞ is not C^2 we do not have at our disposal the curvature tensor, the second variation of arc length, etc., etc. However, the properties (i) - (iii) above are sufficient for having the notion of cut locus and for being able to apply the global theorems (3.5) and (3.16 ii). We will illustrate this by presenting a result of Brittain [Br] (cf. also Katsuda [K]).

(5.16) Brittain. For each $n \geq 2$, $\Delta > 0$ and $V > 0$ there exists a δ such that: Any complete riemannian n-manifold M satisfying $\text{Ric}(M) \geq n-1$, $K_M \leq \Delta$, vol(M) $\geq V$ and diam(M) $\geq \pi - \delta$ is diffeomorphic to S^n.

Proof. Observe that $\text{diam}(M) \leq \pi$ by (4.1) and $K_M \geq (n-1) - (n-2)\Delta$. Suppose M_i is a sequence of manifolds satisfying $\text{Ric}(M_i) \geq n-1$, $K_{M_i} \leq \Delta$, $\text{vol}(M_i) \geq V$ and $\text{diam}(M_i) \uparrow \pi$. By (5.15) we can assume that M_i converges in the Lipschitz metric to a riemannian manifold $M_\infty = (M, g_\infty)$ of class $C^{1,\alpha}$ for all $\alpha \in (0,1)$.

Claim. M_∞ is isometric to $S^n(1)$. To see this observe first that by (i) above $\text{diam}(M_\infty) = \pi$.

Let $p, \bar{p} \in M_\infty$ be at maximal distance. Since (3.16 ii) holds in the limit we can argue as in the proof of (4.3) to get $\text{vol}(B_r^\infty(p)) = \text{vol}(B_r^\infty(\bar{p})) = V_r^1$ for all $r \in [0, \pi]$. In particular $\text{vol}(M_\infty) = \text{vol}(S^n(1))$. Via (3.16 i) this allows us to conclude that all geodesics of length $\leq \pi$ are minimal, i.e. $\text{inj}(M_\infty) = \pi$. Then using (3.16 ii) again we conclude that for each $x \in M_\infty$ there is an $\bar{x} \in M_\infty$ such that $C(x) = \bar{x}$, i.e. M_∞ is a $C^{1,\alpha}$ **wiedersehens manifold**. In the smooth case this implies via a basic inequality of Kazdan [Ka] that such a manifold is isometric to $S^n(1)$ (cf. Berger [Be$_1$]). With more care this inequality can be used also in the $C^{1,\alpha}$ case to conclude that M_∞ is isometric to $S^n(1)$. In fact, for any $x_0 \in M_\infty$, $y_0 \in S^n(1)$ and isometry $I : T_{y_0} S^n(1) \to T_{y_0} M_\infty$ the map defined by

$$F(y) = \begin{cases} \exp_{x_0}(I(\exp_{y_0}^{-1}(y))) &, y \neq -y_0 \\ \bar{x}_0 &, y = -y_0 \end{cases}$$

is an isometry (cf. [Br]).

The theorem clearly follows from the claim.

The following diffeomorphism analogue of (4.7) is a corollary of (5.16) [Br].

(5.17) For each n and $\Delta > 0$ there exists a δ such that: Any complete riemannian n-manifold M satisfying $1 \leq K_M \leq \Delta$ and $\text{diam}(M) \geq \pi - \delta$ is diffeomorphic to S^n.

Proof. In view of (5.16) we only need to see that there is an a priori lower bound for $\text{vol}(M)$, when $1 \leq K_M \leq \Delta$ and, say $\text{diam}(M) > \pi/2$. Let $p, \bar{p} \in M$ be points at maximal distance in M. Then any geodesic loop $\gamma : [0,d] \to M$ at p, i.e. $\gamma(0) = \gamma(d) = p$ has length $L(\gamma) = d > \pi$. To see

this, let γ_1 be a minimal geodesic from $\bar{p}$ to p such that
$\not\prec (\dot\gamma(0), -\dot\gamma_1(\text{diam}(M))) = \alpha \leq \pi/2$ and apply (3.5 B) to the hinge (γ_1,γ,α).
Recall the basic lemma of Klingenberg: If $d(x,y) = d(x,C(x))$ and y is
not conjugate to x along any minimal geodesic from x to y, then there
are exactly two minimal geodesics γ_1, γ_2 from x to y and
$\dot\gamma_1(d(x,y)) = -\dot\gamma_2(d(x,y))$. In our case therefore

$$\text{inj}_p \geq \min\{\pi/2, \pi/\sqrt{\Delta}\},$$

which together with (3.1) gives the desired bound (cf. 3.15 and 3.16).

The injectivity radius estimate by Klingenberg states that $\text{inj}(M) \geq \pi$
for any complete simply connected riemannian manifold with $1/4 < \delta \leq K_M \leq 1$.

The compactness theorem applied to this class of manifolds implies the following well-known pinching theorem of Calabi-Gromoll:

(5.18) For each n there is a δ_n such that any complete simply connected
riemannian manifold M with $\delta_n \leq K_M \leq 1$ is diffeomorphic to S^n.

A corresponding result can be obtained along the same lines for non-simply
connected manifolds M. But then δ_n also depends on the order $|\pi_1(M)|$.
A much better result is known, however:

(5.19) <u>Equivariant pinching theorem.</u> There exists a $1/4 \leq \delta_0 \leq 1$ such
that for any complete simply connected riemannian n-manifold M and any
action $\mu : G \times M \to M$ of a compact Lie group by isometries on M one has:
If $\delta_0 < K_M \leq 1$ then there is an action $\bar\mu : G \times S^n(1) \to S^n(1)$ by isometries
and a diffeomorphism $F : M \to S^n$ such that $\bar\mu(g,F(p)) = F(\mu(g,p))$ for all
$g \in G$, $p \in M$.

The first proof of this theorem [GKR] was based on comparison theory and
the "center of mass"-idea (cf. also [IR]).

We conclude by drawing attention to two further applications of the compactness theorem: The "below 1/4 pinching theorem" by Berger [Be$_2$] and a generalization related to diameter pinching by [D].

References.

[A] U. Abresch, Lower curvature bounds, Toponogov's theorem, and bounded topology, I, Ann. scient. Éc. Norm. Sup. 18 (1985), 563-633; II, to appear.

[Ba] W. Ballmann, Non-positively curved manifolds with rank ≥ 2, Ann. of Math., to appear.

[BBE] W. Ballmann, M. Brin, P. Eberlein, Structure on manifolds with non-positive curvature, I, Ann. of Math., to appear.

[BBS] W. Ballmann, M. Brin, R. Spatzier, Structure on manifolds with non-positive curvature, II, Ann. of Math., to appear.

[Be$_1$] M. Berger, Blaschke's Conjecture for Spheres, Appendix D in A. Besse, Manifolds all of whose Geodesics are Closed, Ergebnisse 93, Springer (1978).

[Be$_2$] M. Berger, Sur les variétés riemanniennes pincées juste au-dessous de 1/4, Ann. Inst. Fourier 33 (1983), 135-150.

[BC] R.L. Bishop, R.J. Crittenden, Geometry of Manifolds, Academic Press (1964).

[Br] D.L. Brittain, A Diameter Pinching Theorem for Positive Ricci Curvature, to appear.

[BS$_1$] K. Burns, R. Spatzier, Classification of a class of topological Tits buildings, Publ. math. I.H.E.S., to appear.

[BS$_2$] K. Burns, R. Spatzier, Manifolds of non positive curvature and their buildings, Publ. math. I.H.E.S., to appear.

[BK] P. Buser, H. Karcher, Gromov's almost flat manifolds, Asterisque 81 (1981).

[CE] J. Cheeger, D.G. Ebin, Comparison Theorems in Riemannian Geometry, North Holland (1975).

[CG$_1$]   J. Cheeger, D. Gromoll, The splitting theorem for manifolds of non-negative Ricci curvature, J. Differential Geometry 6 (1971), 119-128.

[CG$_2$]   J. Cheeger, D. Gromoll, On the structure of complete manifolds of non-negative curvature, Ann. of Math. 96 (1972), 413-443.

[Cg]   S.Y. Cheng, Eigenvalue comparison theorem and its geometric applications, Math.Z. 143 (1975), 289-297.

[Cn]   S.S. Chern, The geometry of G-structures, Bull. Amer. Math. Soc. 72 (1966), 167-219.

[D]   O. Duremeric, A generalization of Berger's almost $\frac{1}{4}$-pinched manifolds theorem, I, Bull. Amer. Math. Soc. 12 (1985), 260-264.

[DK]   D. De Turk, J. Kazdan, Some regularity theorems in riemannian geometry, Ann. Sc. Ec. Norm. Sup. 14 (1981), 249-260.

[GW]   R.E. Greene, H. Wu, Lipschitz convergence of riemannian manifolds, to appear.

[GG]   D. Gromoll, K. Grove, Rigidity of positively curved manifolds with large diameter, Seminar on Differential Geometry, Ann. of Math. Studies, Princeton University Press (1982), 203-207.

[GKM]   D. Gromoll, W. Klingenberg, W.T. Meyer, Riemannsche Geometrie im Grossen, Lecture notes 55, Springer (1968).

[GLP]   M. Gromov, J. Lafontaine, P. Pansu, Structures métrique pour les variétés riemanniennes, Cedic/Fernand Nathan (1981).

[G$_1$]   M. Gromov, Curvature, diameter and Betti numbers, Comment. Math. Helv. 56 (1981), 179-195.

[G$_2$]   M. Gromov, Almost flat manifolds, J. Differential Geometry 13 (1978), 231-242.

[GL$_1$]   M. Gromov, H.B. Lawson, Jr., Spin and scalar curvature in the presence of a fundamental group, Ann. of Math. 111 (1980), 209-230.

[GL$_2$] M. Gromov, H.B. Lawson, Jr., The classification of simply-connected manifolds of positive scalar curvature, Ann. of Math. 111 (1980), 423-434.

[GL$_3$] M. Gromov, H.B. Lawson, Jr., Positive scalar curvature and the Dirac operator on complete riemannian manifolds, Publ. Math. I.A.E.S. 58 (1983), 83-196.

[GK] K. Grove, H. Karcher, How to conjugate C^1-close group actions, Math. Z. 132 (1973), 11-20.

[GKR] K. Grove, H. Karcher, E.A. Ruh, Jacobi fields and Finsler metrics on compact Lie groups with applications to differentiable pinching problems, Math. Ann. 211 (1974), 7-21.

[GS] K. Grove, K. Shiohama, A generalized sphere theorem, Ann. of Math. 106 (1977), 201-211.

[Ha] R.S. Hamilton, Three-manifolds with positive Ricci curvature, J. Differential Geometry 17 (1982), 225-306.

[HK] E. Heintze, H. Karcher, A general comparison theorem with applications to volume estimates for submanifolds, Ann. Sci. Ec. Norm. Sup. 11 (1978), 451-470.

[Hi] N. Hitchin, Harmonic spinors, Adv. in Math. 14 (1974), 1-55.

[IR] H.C. Im-Hof, E.A. Ruh, An equivariant pinching theorem, Comment. Math. Helv. 50 (1975), 389-401.

[I] Y. Itokawa, The topology of certain Riemannian manifolds with positive Ricci curvature, J. Differential Geometry 18 (1983), 151-155.

[JK] J. Jost, H. Karcher, Geometrische Methoden zur Gewinnung von a-priori-Schranken für harmonische Abbildungen, Manuscripta Math. 40 (1982), 27-77.

[K]   A. Katsuda, Gromov's convergence theorem and its application, Nagoya Math. J. 100 (1985), 11-48.

[Ka]   J. Kazdan, An isoperimetric inequality and Wiedersehen manifolds, Ann. of Math. Sudies, Princeton University Press (1982).

[KW]   J. Kazdan, F. Warner, Prescribing curvatures, A.M.S. Proc. of Symp. in Pure Math. 27 (1975), 309-319.

[Kl]   W. Klingenberg, Riemannian Geometry, de Gruyter (1982).

[KN]   S. Kobayashi, K. Nomizu, Foundations of Differential Geometry I, II, Interscience Publishers (1963, 1969).

[L]   A. Lichnerowicz, Spineurs harmoniques, C.r. Acad. Sci. Paris, Sér. A-B, 257 (1963), 7-9.

[M_1] J. Milnor, Morse Theory, Ann. of Math. Studies, Princeton University Press (1963).

[M_2] J. Milnor, A note on curvature and the fundamental group, J. Differential Geometry 2 (1968), 1-7.

[Pe_1] S. Peters, Cheeger's finiteness theorem for diffeomorphism classes of riemannian manifolds, J. Reine Angew. Math. 394 (1984), 77-82.

[Pe_2] S. Peters, Convergence of Riemannian manifolds, to appear.

[Po]   W.A. Poor, Some results on nonnegatively curved manifolds, J. Differential Geometry 9 (1974), 583-600.

[R]   J. Rosenberg, C^*-algebras, positive scalar curvature and the Novikov conjecture I, Publ. Math. I.H.E.S. 58 (1983), 197-212, II, III to appear.

[SY_1] R. Schoen, S.T. Yau, On the structure of manifolds with positive scalar curvature, Manuscripta Math. 28 (1979), 159-183.

[SY₂]   R. Shoen, S.T. Yau, Complete three dimensional manifolds with positive Ricci curvature and scalar curvature, Seminar on Differenial Geometry, Ann. of Math. Studies, Princeton University Press (1982), 209-227.

[S]   K. Shiohama, A sphere theorem for manifolds of positive Ricci curvature, Trans. Amer. Math. Soc. 275 (1983), 811-819.

COMPLEX DIFFERENTIAL GEOMETRY

Robert E. Greene

1. Complex manifolds

Let $\mathbf{C}$ = the complex numbers and $\mathbf{C}^n = \mathbf{C} \times \cdots \times \mathbf{C}$ (n factors). $\mathbf{C}^n$ will be referred to as n-dimensional complex euclidean space. Of course, $\mathbf{C}^n$ is topologically $2n$-dimensional; more specifically, $\mathbf{C}^n$ can and will be identified homeomorphically with $\mathbf{R}^{2n}$ as follows: If $(z_1,\ldots,z_n) \in \mathbf{C}_n$ and if $z_j = x_j + iy_j, x_j, y_j \in \mathbf{R}$, $j = 1,\ldots n$ then $(z_1,\ldots,z_n)$ is to be identified with $(x_1,y_1,\ldots,x_n,y_n) \in \mathbf{R}^{2n}$. That is, $(z_1,\ldots,z_n)$ is identified with (Re z_1, Im z_1,...,Re z_n, Im z_n), where Re and Im denote real and imaginary parts, respectively.

DEFINITION. A function $f : D \to \mathbf{C}$ defined on an open subset D of $\mathbf{C}^n$ is *holomorphic* if f considered as a function from $D \subset \mathbf{R}^{2n}$ to $\mathbf{R}^2$ is C^∞ and satisfies the Cauchy-Riemann equations:

$$\frac{\partial(\text{Re } f)}{\partial x_j} = \frac{\partial(\text{Im } f)}{\partial y_j} \quad \text{and} \quad \frac{\partial(\text{Re } f)}{\partial y_j} = -\frac{\partial(\text{Im } f)}{\partial x_j}$$

for all $j = 1,\ldots,n$. A mapping $f : D \to \mathbf{C}^m$ is *holomorphic* if each component of f is holomorphic; that is, if $f_1,\ldots,f_m$ are holomorphic, where $f_j : D \to \mathbf{C}, j = 1,\ldots,m$ are defined by $f(z) = (f_1(z),\ldots,f_m(z)) \in \mathbf{C}^m$ for $z \in D$.

Note that if $D = \cup_{\lambda \in \Lambda} D_\lambda$ with each D_λ open, then $f : D \to \mathbf{C}^m$ is holomorphic if and only if $f \mid D_\lambda$ is holomorphic for every λ.

Multiplication by i, $(z_1,\ldots,z_n) \to (iz_1,\ldots,iz_n)$, defines a mapping from $\mathbf{C}^n$ to $\mathbf{C}^n$, whose square is multiplication by -1. The induced mapping from $\mathbf{R}^{2n}$ to $\mathbf{R}^{2n}$, which will be denoted by J, is a real linear endomorphism, whose composition with itself is again multiplication by -1. The following lemma shows that the endomorphism J can be used to characterize holomorphic mappings:

LEMMA 1. *A mapping* $f: D \to \mathbf{C}^m$, $D^{open} \subset \mathbf{C}^n$, *which is* C^∞ *when considered as a function from* $D \subset \mathbf{R}^{2n}$ *to* R^{2m} *is holomorphic if and only if* $f_* J_{\mathbf{R}^{2n}} = J_{\mathbf{R}^{2m}} f_*$. *Here* f_* *is the real Jacobian of* f.

Proof. Write $f(z) = (f_1(z),...,f_m(z))$, $f_j(z) \in \mathbf{C}$ for $j = 1,...,m$. In matrix form relative to the standard bases of $\mathbf{R}^{2n}$ and $\mathbf{R}^{2m}$,

$$f_* = \begin{pmatrix} \frac{\partial(\mathrm{Re}f_1)}{\partial x_1} & \frac{\partial(\mathrm{Re}f_2)}{\partial y_1} & \cdots & \frac{\partial(\mathrm{Re}f_m)}{\partial x_n} & \frac{\partial(\mathrm{Re}f_m)}{\partial y_n} \\ \frac{\partial(\mathrm{Im}f_1)}{\partial x_1} & \frac{\partial(\mathrm{Im}f_1)}{\partial y_1} & \cdots & \frac{\partial(\mathrm{Im}f_m)}{\partial x_n} & \frac{\partial(\mathrm{Im}f_m)}{\partial y_n} \\ \vdots & \vdots & & \vdots & \vdots \\ \frac{\partial(\mathrm{Re}f_m)}{\partial x_1} & \frac{\partial(\mathrm{Re}f_m)}{\partial y_1} & \cdots & \frac{\partial(\mathrm{Re}f_m)}{\partial x_n} & \frac{\partial(\mathrm{Re}f_m)}{\partial y_n} \\ \frac{\partial(\mathrm{Im}f_m)}{\partial x_1} & \frac{\partial(\mathrm{Im}f_m)}{\partial y_1} & \cdots & \frac{\partial(\mathrm{Im}f_m)}{\partial x_n} & \frac{\partial(\mathrm{Im}f_m)}{\partial y_n} \end{pmatrix}$$

$$J_{\mathbf{R}^{2n}} = \begin{pmatrix} \begin{matrix} 0 & -1 \\ 1 & 0 \end{matrix} & & & 0 \\ & \begin{matrix} 0 & -1 \\ 1 & 0 \end{matrix} & & \\ & & \ddots & \\ 0 & & & \begin{matrix} 0 & -1 \\ 1 & 0 \end{matrix} \end{pmatrix}$$

$$J_{\mathbf{R}^{2n}} = \begin{pmatrix} 0 & -1 & & & & & & \\ 1 & 0 & & & & & & \\ & & & & & & 0 & \\ & & 0 & -1 & & & & \\ & & 1 & 0 & & & & \\ & & & & \ddots & & & \\ & & 0 & & & & 0 & -1 \\ & & & & & & 1 & 0 \end{pmatrix}$$

Straightforward matrix multiplication shows that $f_* J = J f_*$ if and only if

$$\frac{\partial (\operatorname{Re} f_k)}{\partial x_j} = \frac{\partial (\operatorname{Im} f_k)}{\partial y_j} \quad \text{and} \quad \frac{\partial (\operatorname{Im} f_k)}{\partial x_j} = - \frac{\partial (\operatorname{Re} f_k)}{\partial y_j}$$

for all $k = 1,\ldots,m$ and $j = 1,\ldots,n$.

Lemma 1 implies immediately the following corollaries:

COROLLARY. *If $f : D \to \mathbf{C}^m$ and $g : D_1 \to \mathbf{C}^k$ are holomorphic mappings and if $f(D) \subset D_1^{\text{open}} \subset \mathbf{C}^m$, then $g \circ f : D \to \mathbf{C}^k$ is holomorphic.*

COROLLARY. *If $f : D \to \mathbf{C}^n$, $D^{\text{open}} \subset \mathbf{C}^n$, is holomorphic and is a diffeomorphism onto its image when considered as a mapping from $D \subset \mathbf{R}^{2n}$ to $\mathbf{R}^{2n}$ then $f^{-1} : f(D) \to \mathbf{C}^n$ is holomorphic. (Note that in this case $f(D)$ is necessarily open in $\mathbf{C}^n$.)*

DEFINITION. A *complex structure* on a C^∞ manifold M of even dimension $2n$ is a maximal collection of C^∞ charts indexed by a set Λ:

$$\{(\psi_\lambda, U_\lambda) : \lambda \in \Lambda, \psi_\lambda : U_\lambda \to R^{2n}\}$$

having $\bigcup_{\lambda \in \Lambda} U_\lambda = M$ and satisfying the following condition (holomorphic transition) for all $\lambda, \mu \in \Lambda$,

$$\psi_\lambda \psi_\mu^{-1} : \psi_\mu(U_\lambda \cap U_\mu) \to \psi_\lambda(U_\lambda \cap U_\mu) \subset \mathbf{R}^{2n}$$

is a holomorphic function considered as a function from an open subset of $\mathbf{C}^n$ to $\mathbf{C}^n$.

A *complex manifold* is a paracompact C^∞ manifold M together with a complex structure. The number n is the *complex dimension* of M. The coordinate charts of the complex structure of a complex manifold are called *holomorphic coordinate charts* or *holomorphic coordinate systems*.

Note that an open subset of a complex manifold is itself a complex manifold in a standard fashion.

LEMMA 2. *Any collection of C^∞ charts on a paracompact C^∞ manifold M which cover M and which satisfy the holomorphic transition condition determine one and only one complex structure on M; i.e. any such collection is contained in a unique maximal such collection.*

The proof of this lemma follows exactly the pattern (using the corollaries of Lemma 1) of the standard proof of the corresponding result for real manifolds.

Let M be a complex manifold and p be a point of M. Denote the (real) tangent space of M at p by M_p. An endomorphism $J_p : M_p \to M_p$ of the real vector space M_p with $J_p^2 =$ multiplication by -1 can be defined as follows: Choose a holomorphic coordinate system $\psi : U \to \mathbf{R}^{2n}$ defined in a neighborhood U of p. Then $\psi_{*p} : M_p \to \mathbf{R}^{2n}$ is an isomorphism (the tangent space to $\mathbf{R}^{2n}$ at $\psi(p)$ being identified with $\mathbf{R}^{2n}$). J_p is then defined to be $\psi_{*p}^{-1} J_{\mathbf{R}^{2n}} \psi_{*p}$. Lemma 2 implies immediately that J_p thus defined does not depend on the choice of ψ: if ϕ is a second holomorphic coordinate system around p then at p

$$\phi_*^{-1} J_{\mathbf{R}^{2n}} \phi_* = \phi_*^{-1} (\psi\phi^{-1})_*^{-1} J_{\mathbf{R}^{2n}} (\psi\phi^{-1})_* \phi_*$$

because $(\psi\phi^{-1})_*$ commutes with $J_{\mathbf{R}^{2n}}$ by Lemma 2 and the holomorphicity of $\psi\phi^{-1}$. Hence

$$\phi_*^{-1} J_{\mathbf{R}^{2n}} \phi_* = \phi_*^{-1} \phi_* \psi_*^{-1} J_{\mathbf{R}^{2n}} \psi_* \phi_*^{-1} \phi_* = \psi_*^{-1} J_{\mathbf{R}^{2n}} \psi_*.$$

Thus J_p is independent of holomorphic coordinate choice; that $J_p^2 =$ multiplication by

-1 is apparent from the fact that $J_{\mathbf{R}^{2n}}^2 = $ multiplication by -1.

Since holomorphic coordinate systems are always C^∞, $p \to J_p$ considered as a (1,1) tensor on M is C^∞. Note that this C^∞ tensor completely determines the complex structure on M since a C^∞ coordinate system $\psi : U \to \mathbf{R}^{2n}$ is a holomorphic coordinate system if and only if $\psi_{*p} J_p = J_{\mathbf{R}^{2n}} \psi_{*p}$ for all $p \in U$.

DEFINITION. A C^∞ mapping $f : M \to M'$ from one complex manifold M to another M' is *holomorphic* if $f_* J_M = J_{M'} f_*$.

Lemma 2 shows that this definition is equivalent to the requirement that f be holomorphic when expressed (locally) in holomorphic coordinate systems on M and M'. The definition as given has the advantage of avoiding local coordinate expressions.

Note that if M' is a complex manifold and M a C^∞ real submanifold then M has at most one complex structure such that the injection $i : M \to M'$ is holomorphic since at most one J_M can satisfy $i_* J_M = J_{M'} i_*$, i_* being injective. Of course such a complex structure on M may fail to exist. If one (and necessarily only one) such does exist, then M is said to be a *complex submanifold* of M'.

Since the (1,1) tensor J with $J^2 = $ multiplication by -1 on a complex manifold determines the complex structure, it is reasonable to consider such tensors independent of any complex structure:

DEFINITION. Let M be a C^∞ manifold. An *almost complex structure* on M is a C^∞ (1,1) tensor J on M such that, for every $p \in M$, $J_p^2 : M_p \to M_p$ is multiplication by -1.

LEMMA 3. *A real finite dimensional vector space which admits an endomorphism $J : V \to V$ satisfying $J^2 = $ multiplication by -1 is necessarily even dimensional.*

Proof. Let $X, Y \to g(X,Y)$ be any positive definite inner product on V. Then

$(X,Y) \to H(X,Y)$ defined by $H(X,Y) = g(X,Y) + g(JX,JY)$ is a positive definite inner product on V relative to which J acts as an isometry. In particular it follows that any J-invariant subspace has a J-invariant complementary subspace.

Now note that if $X_1 \in V$ and $X_1 \neq 0$ then X_1 and JX_1 are real linearly independent since if $JX_1 = \alpha X_1$, $\alpha \in \mathbf{R}$, then $X_1 = -J^2 X_1 = -\alpha JX_1 = -\alpha^2 X_1$ or $\alpha^2 = -1$, which is impossible for $\alpha \in \mathbf{R}$. Hence X_1 and JX_1 span a 2-dimensional subspace, which is obviously J-invariant. This subspace has then a J-invariant complement, say V_1, of dimension equal to two less than the dimension of V. The lemma now follows by an obvious induction on dimension argument.

REMARK. The existence of a J-invariant complement for every J-invariant subspace is of course a special case of the complete reducibility of representations of finite groups.

Lemma 3 shows that any manifold admitting an almost complex structure must be even dimensional.

LEMMA 4. *Let M be a paracompact (even dimensional) C^∞ manifold with an almost complex structure J_M. In order that J be the almost complex structure on M associated to complex structure on M, it is necessary and sufficient that for each point p of M there exists a C^∞ coordinate chart $\psi : U \to \mathbf{R}^{2n}$, $p \in U$, such that $\psi_* J_M = J_{\mathbf{R}^{2n}} \psi_*$ everywhere on U.*

Proof. The necessity of the condition is immediate. To prove the sufficiency, note that if $\psi : U_1 \to \mathbf{R}^{2n}$ and $\phi : U_2 \to \mathbf{R}^{2n}$ satisfy $\psi_* J_M = J_{\mathbf{R}^{2n}} \psi_*$ everywhere on U_1 and $\phi_* J_M = J_{\mathbf{R}^{2n}} \phi_*$ everywhere on U_2 then $\psi \phi^{-1} : \phi(U_1 \cap U_2) \to \psi(U_1 \cap U_2)$ satisfies

$$(\psi \phi^{-1})_* J_{\mathbf{R}^{2n}} = \psi_* \phi_*^{-1} J_{\mathbf{R}^{2n}} = \psi_* J_M \phi_*^{-1} = J_{\mathbf{R}^{2n}} \psi_* \phi_*^{-1} = J_{\mathbf{R}^{2n}} (\psi \phi^{-1})_*$$

everywhere on $\phi(U_1 \cap U_2)$. By Lemma 1, $\psi \phi^{-1}$ is holomorphic on the open set

$\phi(U_1 \cap U_2)$, and Lemma 2 now implies the desired conclusion.

It is possible to find conditions expressed directly in terms of the almost complex structure tensor J which are necessary and sufficient for J to arise from a complex structure. These conditions are discussed in the Appendix Z.

2. Basic Examples

Two complex manifolds M and N are *biholomorphic* if there is a (real) diffeomorphism $F : M \to N$ that is holomorphic. It follows from the results in §1 that F^{-1} is then also holomorphic. From this, it then follows that the relation of being biholomorphic is an equivalence relation. Two complex manifolds that are biholomorphic are the same object as far as the purposes of complex manifold theory go.

The most fundamental examples for the theory are the following:

(i) $\mathbf{C}^n$. This is $\mathbf{R}^{2n}$ with the complex structure J determined by

$$\mathbf{R}^{2n} = \{(x_1, y_1, \ldots, x_n, y_n)\}$$

$$J\left(\frac{\partial}{\partial x_i}\right) = \frac{\partial}{\partial y_i}, \quad J\left(\frac{\partial}{\partial y_i}\right) = -\frac{\partial}{\partial x_i} \quad i = 1, 2, \ldots, n.$$

The holomorphic coordinate system $(x_1 + \sqrt{-1}\, y_1, \ldots, x_n + \sqrt{-1}\, y_n)$ determines a complex manifold structure.

(ii) $B^n = \left\{(z_1, \ldots, z_n) \in \mathbf{C}^n \mid \sum_{j=1}^{n} |z_j|^2 < 1\right\},$

the unit ball in $\mathbf{C}^n$. Since this is an open subset of $\mathbf{C}^n$, it inherits a complex manifold structure automatically. It is important to note that $\mathbf{C}^n$ and B^n are *not* biholomorphic even though they are real diffeomorphic. To see this, note that a holomorphic mapping $F : \mathbf{C}^n \to B^n$, $F = (F_1, \ldots, F_n)$, $F_j : \mathbf{C}^k \to \mathbf{C}$ is necessarily constant because Liouville's Theorem immediately implies that each F_j is constant. (In detail: $z \to F_j(a + bz)$, $a, b \in \mathbf{C}$, $z \in \mathbf{C}$ is constant so F_j is itself constant.)

(iii) Variations of B^n. If S is a C^∞ submanifold that is C^1 close to $\{(x_1,y_1,...,x_n,y_n) \mid \Sigma x_i^2 + \Sigma y_i^2 = 1\}$, then the interior U_s of S is an open subset of $\mathbf{C}^n$ that is real diffeomorphic to B^n. If $n = 1$, then U_s is biholomorphic to B^1, by the Riemann Mapping Theorem. But if $n \geq 2$, U_s is generically *not* biholomorphic to B^n. The U_s so obtained in fact form a collection of complex manifolds that in a suitable sense give an infinite dimensional family of biholomorphic equivalence classes.

(iv) $P_n\mathbf{C}$, complex projective space of (complex) dimension n. The notation $\mathbf{C}P^n$ is also used. To define $P_n\mathbf{C}$, define an equivalence relation on $\mathbf{C}^{n+1} - \{(0,...,0)\}$ as follows:

$$(z_1,...,z_{n+1}) \sim (w_1,...,w_{n+1})$$

if and only if $\exists \lambda \in \mathbf{C}$ such that $z_j = \lambda w_j$ for each $j = 1,2,...,n+1$. (Exercise: Check that this is an equivalence relation.) As a set, $P_n\mathbf{C}$ is the set of all equivalence classes of $\mathbf{C}^{n+1} - \{(0,...,0)\}/\sim$. Define a topology on $P_n\mathbf{C}$ by declaring a set $U \subset P_n\mathbf{C}$ to be open if and only if

$$\{(z_1,...,z_{n+1}) \in \mathbf{C}^{n+1} - \{(0,...,0)\} \mid [(z_1,...,z_{n+1})] \in U\}$$

is open in $\mathbf{C}^{n+1}$, where [g] denotes the equivalence class of $(z_1,...,z_{n+1})$. (This is the topology usually called the quotient topology relative to the equivalence relation.) $P_n\mathbf{C}$ is compact because the map

$$\{(z_1,...,z_{n+1}) \mid \Sigma|z_j|^2 = 1\} \to P_n\mathbf{C}$$

obtained by taking equivalence classes is surjective (and continuous by definition).

To define a complex structure on $P_n\mathbf{C}$, we exhibit coordinate systems as follows: For each j, set $U_j = \{[(z_1,...,z_{n+1})] \mid z_j \neq 0\}$. Clearly $\cup U_j = P_n\mathbf{C}$. Define $F_j : U_j \to \mathbf{C}^n$ by

$$[(z_1,...,z_{n+1})] \to (z_1/z_j,...,z_l/z_j,...,z_n/z_j), \quad \ell \neq j.$$

The transition map from $F_j(U_j)$ to $F_k(U_k)$ defined on $F_j(U_j \cap U_k)$ is easily seen to be holomorphic: it is essentially multiplication by z_j/z_k. (Exercise: Check details of this.) Thus the $F_j : U_j \to \mathbf{C}^n$ maps define a complex manifold structure on $P_n\mathbf{C}$.

(v) Submanifolds (of $P_n\mathbf{C}$): The definition of (complex) submanifold of a complex manifold runs parallel with the real theory: A subset N of a complex manifold M is an (embedded) submanifold if (1) N is a C^∞ real embedded submanifold of M and (2) T_pN, all $p \in N$, is a J invariant subspace of T_pM (T_pN = real tangent space) of N at p.

This definition condition (2) can be checked to be equivalent to the idea that in a neighborhood in M of each point $q \in N$, $N \cap$ neighborhood is a slice in some holomorphic coordinate system, i.e.,

$$N \cap \text{nbhd} = \{(z_1,...,z_n) \mid z_j = z_{j+1} = \cdots = z_n = 0\}$$

in some (holomorphic) coordinates $(z_1,...,z_n)$.

A compact submanifold of $P_n\mathbf{C}$ is called a (compact) algebraic variety. The fact that a compact complex submanifold of $P_n\mathbf{C}$ is an algebraic variety in the usual sense (of being the common zeroes of a set of homogeneous polynomials) is true, but hard to prove. A converse statement is, however, relatively easy.

If $P(z_1,...,z_{n+1})$ is a homogeneous polynomial, then, for $\lambda \neq 0$, $P(\lambda z_1,...,\lambda z_{n+1}) = 0$ if and only if $P(z_1,...,z_{n+1}) = 0$ since

$$P(\lambda z_1,...,\lambda z_{n+1}) = \lambda^d P(z_1,...,z_n)$$

where d = degree (of homogeneity) of P. Thus it makes sense to refer to P vanishing (or not) at a point of $P_n\mathbf{C}$. If $P_1,...,P_k$ is a (finite) set of polynomials that are homogeneous, set $V(P_1,...,P_k)$ = the points of $P_n\mathbf{C}$ at which all the $P_1,...,P_k$ vanish. The set $V(P_1,...,P_k)$ need not be a real submanifold: it could have singularities. But if it is a real submanifold, then it is in fact necessarily a complex submanifold. Rather than considering this fact in full generality, we illustrate with a concrete example.

Set
$$V = V(z_1^2 + z_2^2 + z_3^2) \subset P_2\mathbf{C} .$$

Consider $V \cap U_j$, $j \in \{1,2,3\}$ where $U_j = \{[(z_1,z_2,z_3)] | z_j \neq 0\}$ as before, with $j = 3$, say. The coordinates on U_3 are (z_1,z_2) with $(z_1,z_2) \leftrightarrow [(z_1,z_2,1)] \in P_2\mathbf{C}$. The polynomial vanishes at $[(z_1,z_2,1)]$ if and only if $z_1^2 + z_2^2 + 1 = 0$. So $V \cap U_3$ corresponds in (z_1,z_2) coordinates to

$$\{(z_1,z_2) \mid z_1^2 + z_2^2 + 1 = 0\} .$$

This is a complex submanifold in $\mathbf{C}^2$ because, near (w_1,w_2), $w_2 \neq 0$, with $w_1^2 + w_2^2 + 1 = 0$, the functions

$$(z_1, z_1^2 + z_2^2 + 1)$$

form a holomorphic coordinate system and $\{(z_1,z_2) \mid z_1^2 + z_2^2 + 1 = 0\}$ is obviously a coordinate slice in this coordinate system (what if $w_2 = 0$? Exercise). So, checking similarly on $V \cap U_1$ and $V \cap U_2$, we see that V is a complex submanifold of $P_2\mathbf{C}$.

It is important to realize that the analogue of Whitney's Embedding Theorem does not hold in the complex case. First of all, a connected, compact complex submanifold of $\mathbf{C}^n$ must be a point. (Proof outline: $z_j | N$ is holomorphic if N is a complex submanifold of $\mathbf{C}^n$. If N is compact and connected, the Maximum Modulus Principle implies z_j is constant. Think for yourself about why the Maximum Modulus Principle applies in several variables!) So $P_n\mathbf{C}$ is *not* realizable as a submanifold of $\mathbf{C}^n$.

You might be inclined to hope that $P_n\mathbf{C}$ would play some sort of universal embedding theorem role. But this does not work, either. There are compact complex manifolds that are not submanifolds (in the complex sense) of any $P_n\mathbf{C}$. It is not trivial to see why, however. See §3.

3. Hermitian and Kähler Metrics

DEFINITION. A C^∞ Riemannian metric g on a complex manifold M is an

Hermitian metric if for each $p \in M$ and each pair of tangent vectors $X,Y \in M_p$, the tangent space of M at p,

$$g(X,Y) = g(JX,JY).$$

Every complex manifold admits an Hermitian metric. To verify this fact, note that, since a complex manifold M is paracompact by definition, it necessarily admits a C^∞ Riemannian metric g_1. If g is defined by $g(X,Y) = g_1(X,Y) + g_1(JX,JY)$, then g is clearly an Hermitian metric.

The following lemma describes the pointwise structure of Hermitian metrics:

LEMMA 1. *Let V be a real vector space and J an endomorphism of V satisfying $J^2 =$ multiplication by -1. Suppose that g is a positive definite inner product on V satisfying $g(X,Y) = g(JX,JY)$ for all $X,Y \in V$. Then*

(a) $g(X,JY) = -g(JX,Y)$.

(b) *there exists an orthonormal basis for V of the form $X_1, JX_1, X_2, JX_2, \ldots, X_n, JX_n$.*

Proof. To prove (a), note that $g(X,JY) = g(JX,J(JY))$ while $g(JX,J(JY)) = g(JX, -Y) = -g(JX,Y)$.

To prove (b), one follows the method used to prove Lemma 3 of §1: Let X_1 be a unit vector in V. Then JX_1 is a unit vector since $g(JX_1,JX_1) = g(X_1,X_1)$. Also $g(X_1,JX) = -g(JX_1,X_1)$ by part (a) so $g(X_1,JX_1) = 0$. The subspace generated by X_1 and JX_1 is J-invariant, and hence its orthogonal complement is also J-invariant. The desired conclusion now follows by induction on the dimension of V.

If g is an Hermitian metric, then the 2-tensor ω defined by $\omega(X,Y) = g(JX,Y)$ is antisymmetric by part (a) of Lemma 1.

DEFINITION. If g is an Hermitian metric, the 2-form ω defined by $\omega(X,Y) = g(JX,Y)$ is the *Kähler form* of g.

LEMMA 2. *If g is an Hermitian metric, then $\omega \wedge \cdots \wedge \omega$ (n factors, $n =$ the complex dimension of M) is nowhere zero.*

Proof. Given $p \in M$, let $X_1, JX_1, \ldots, X_n, JX_n$ be an orthonormal basis relative to g for M_p. Such a basis exists by Lemma 1b). Then

$$\omega(X_i, JX_i) = g(JX_i, JX_i) = 1$$

while

$$\omega(X_j, JX_i) = g(JX_j, JX_i) = 0 \text{ if } i \neq j.$$

Also

$$\omega(X_i, X_j) = \omega(JX_i, JX_j) = 0 \text{ for all } i, j.$$

Then

$$(\omega \wedge \cdots \wedge \omega)(X_1, JX_1, \ldots, X_n, JX_n) = n! \, \omega(X_1, JX_1) \cdot \cdots \cdot \omega(X_n, JX_n) = n!$$

In particular, $(\omega \wedge \cdots \wedge \omega) \neq 0$ at p.

PROPOSITION 1. *A complex manifold is orientable.*

Proof. Since any complex manifold admits an Hermitian metric, Lemma 2 shows that there exists a nowhere vanishing $2n$ form on any complex manifold of real dimension $2n$.

Actually, it can be shown that a complex manifold is orientable without introducing any metric concepts. In fact, a holomorphic mapping from a domain in $\mathbf{C}^n$ to $\mathbf{C}^n$ which is a real diffeomorphism onto its image is necessarily orientation preserving on the underlying $\mathbf{R}^{2n}$ since its (real) Jacobian determinant is positive (by a calculation with determinants using the Cauchy-Riemann equations). Thus a covering by coordinate systems all of whose transition mappings have positive Jacobian determinant is given directly by the complex structure.

DEFINITION. An Hermitian metric g on a complex manifold is a *Kähler metric* if the Kähler form ω associated to g is closed, i.e. $d\omega = 0$.

The condition $d\omega = 0$ implies a surprisingly close relationship between the metric g and the complex structure of M. One aspect of this relationship is expressed in the following proposition.

PROPOSITION 2. *Let M be a complex manifold and g an Hermitian metric on M. Then the following conditions on g are equivalent:*

(a) *g is a Kähler metric*

(b) *If D is the Riemannian covariant differentiation associated to g, then $DJ = 0$.*

Proof that (b) implies (a): Since $Dg = 0$ by definition of g and $\omega(X,Y) = g(JX,Y)$ for all X, Y, the vanishing of DJ implies that of $D\omega$:

$$D_Z\omega(X,Y) = Z\,\omega(X,Y) - \omega(D_Z X, Y) - \omega(X, D_Z Y)$$

$$= Zg(JX,Y) - g(J(D_Z X),Y) - g(JX, D_Z Y)$$

$$= (D_Z g)(JX,Y) + g(D_Z(JX),Y) + g(JX, D_Z Y) - g(J(D_Z X),Y) - g(JX, D_Z Y)$$

$$= 0 + g((D_Z J)X,Y) + g(J(D_Z X),Y) + g(JX, D_Z Y) - g(J(D_Z X),Y) - g(JX, D_Z Y)$$

$$= g(D_Z J)X,Y) = 0 \text{ if } DJ = 0.$$

Now for any form ω, the formula

$$d\omega = \sum_i du^i \wedge D_{\partial/\partial u^i}\,\omega$$

holds, where (u^i) is a real C^∞ local coordinate system. In particular, if $D\omega = 0$ then $d\omega = 0$.

Proof that (a) implies (b): It suffices to establish the following formula for an arbitrary Hermitian metric g and its Kähler form ω:

$$g((D_Z J)X,Y) = \frac{1}{2} d\omega(X,Y,Z) - \frac{1}{2} d\omega(JX,JY,Z).$$

For then the vanishing of $d\omega$ implies that $g(D_Z J)X,Y) = 0$ for all X and Y and hence that $D_Z J = 0$. To establish the formula, note first that, all terms being tensors, the formula need be verified only in the case that X, Y, and Z are linear combinations with constant coefficients of the standard vector fields associated to a C^∞ local coordinate system. In fact, this coordinate system may be taken to be the real and imaginary parts of a holomorphic one, say $(x_1,y_1,x_2,y_2,\ldots,x_n,y_n)$. Note that if X, Y, and Z are (locally) such constant coefficient linear combinations of $\partial/\partial x_i$ and $\partial/\partial y_i$ then so are JX, JY, and JZ so that the Lie brackets of any two of the six vector fields X, Y, Z, JX, JY, JZ all vanish.

Then

$$g((D_Z J)X,Y) = g(D_Z(JX) - J(D_Z X),Y)$$

$$= g(D_Z(JX),Y) - g(J(D_Z X),Y) = g(D_Z(JX),Y) + g(D_Z X,JY)$$

$$= \frac{1}{2}[(JX)g(Y,Z) + Zg(JX,Y) - Yg(JX,Z)]$$

$$+ \frac{1}{2}[Zg(X,JY) + Xg(JY,Z) - (JY)g(X,Z)];$$

and on the other hand

$$d\omega(X,Y,Z) = X\omega(Y,Z) + Z\omega(X,Y) - Y\omega(X,Z)$$

$$= Xg(JY,Z) + Zg(JX,T) - Yg(JX,Z)$$

and

$$d\omega(JX,JY,Z) = (JX)\omega(JY,Z) + Z\omega(JX,JY) - (JY)\omega(JX,Z)$$

$$= -(JX)g(Y,Z) - Zg(X,JY) + (JY)g(X,Z).$$

The desired formula thus follows.

Proposition 2 implies that the tensor J is invariant under parallel translation relative to a Kähler metric; in particular, the value J_p of J at a single point $p \in M$ together with the Kähler metric on M determines J everywhere on M (assuming that M is connected) and thus the complex structure of M.

If g is a Kähler metric, the Kähler form ω, being closed, determines a class in deRham 2-cohomology. Because $\omega \wedge \cdots \wedge \omega$ (n times) is nowhere vanishing, and hence a nonvanishing volume form multiple, ω cannot be exact. (Detail: If $\omega = d\theta$, then $\omega \wedge \cdots \wedge \omega$ (n times) $= d(\theta \wedge \omega \cdots \wedge \omega)$, $n - 1$ ω's. So

$$0 = \int d(\theta \wedge \omega \wedge \cdots \wedge \omega) = \int \omega \wedge \cdots \wedge \omega \neq 0 ,$$

a contradiction.) Thus the cohomology class of ω is nonzero. Hence any Kähler manifold has nonvanishing 2-cohomology in the deRham sense. It can be seen that $M = S^{2p+1} \times S^{2q+1}$ admits a complex structure, all $p,q \geq 1$ (cf. [2]). But then M has zero 2-cohomology and hence does not admit Kähler metrics (see §10), so such an M cannot be realized as a submanifold of $P_n\mathbf{C}$.

4. Complexification of the Tangent and Cotangent Spaces

DEFINITION. Let V be a real vector space. The *complexification* of V, to be denoted $V^\mathbf{C}$, is the complex vector space consisting of all ordered pairs $(v,w), v, w \in V$ with operations defined by

$$(v_1, w_1) + (v_2, w_2) = (v_1 + w_1, v_2 + w_2) .$$

$$(\alpha + i\beta)(v, w) = (\alpha v - \beta w, \alpha w + \beta v) , \quad \alpha, \beta \in \mathbf{R} .$$

LEMMA 1. *If V is a real vector space of dimension n and $v_1, \ldots, v_n$ is a basis for V, then*

(a) *$V^\mathbf{C}$ has complex dimension n and $(v_1, 0) \cdots (v_n, 0)$ is a (complex) basis for $V^\mathbf{C}$*

(b) *$V^\mathbf{C}$ considered as a real vector space has dimension $2n$ and $(v_1, 0), \ldots, (v_n, 0)$, $(0, v_1), \ldots, (0, v_n)$ is a (real) basis for $V^\mathbf{C}$.*

Proof. As an example, the complex linear independence of $(v_1,0),\ldots,(v_n,0)$ will be verified, the other verifications being left to the reader: If for $\alpha_\ell, \beta_\ell \in R$, $\ell = 1,\ldots,n$,

$$\sum_{\ell=1}^{n} (\alpha_\ell + i\beta_\ell)(v_\ell,0) = (0,0)$$

then

$$(0,0) = \sum_{\ell=1}^{n} (\alpha_\ell v_\ell, \beta_\ell v_\ell) = \left(\sum_{\ell}^{n} \alpha_\ell v_\ell, \sum_{\ell=1}^{n} \beta_\ell v_\ell \right)$$

or $\sum_{\ell=1}^{n} \alpha_\ell v_\ell = 0$ and $\sum_{\ell=1}^{n} \beta_\ell v_\ell = 0$. Hence $\alpha_\ell = 0$ for all ℓ and $\beta_\ell = 0$ for all ℓ by the real linear independence of the $v_1,\ldots,v_n$.

Since in V^C $i(w,0) = (0,w)$ for any $w \in V$, it is reasonable to introduce the notation $v + iw$ for the element (v,w) of V^C.

DEFINITION. Let M be a complex manifold of complex dimension n, p be a point of M, and $(z_1,\ldots,z_n) = (x_1,y_1,\ldots,x_n,y_n)$ be a holomorphic coordinate system defined in a neighborhood of p. Then $\partial/\partial z_i|_p$ and $\partial/\partial \bar{z}_i|_p$, $i = 1,\ldots,n$, are the elements of M_p^C given by

$$\left.\frac{\partial}{\partial z_i}\right|_p = \frac{1}{2}\left(\left.\frac{\partial}{\partial x_i}\right|_p - i\left.\frac{\partial}{\partial y_i}\right|_p\right)$$

$$\left.\frac{\partial}{\partial \bar{z}_i}\right|_p = \frac{1}{2}\left(\left.\frac{\partial}{\partial x_i}\right|_p + i\left.\frac{\partial}{\partial y_i}\right|_p\right).$$

When the point p is clear from the context the abbreviated notations $\partial/\partial z_i$, $\partial/\partial \bar{z}_i$ will be used.

Since

$$\frac{\partial}{\partial x_i} = \frac{\partial}{\partial z_i} + \frac{\partial}{\partial \bar{z}_i}$$

and

$$\frac{\partial}{\partial y_1} = -i\left(\frac{\partial}{\partial z_i} - \frac{\partial}{\partial \bar{z}_i}\right)$$

it follows from Lemma 1 that $\partial/\partial z_1,\ldots,\partial/\partial z_n$, $\partial/\partial \bar{z}_1,\ldots,\partial/\partial \bar{z}_1$, being thus a spanning set of $2n$ elements, are a complex basis for M_p^C.

Lemma 2. *The complex subspace of M_p^C spanned by $\partial/\partial z_1,\ldots,\partial/\partial z_n$ is independent of the choice of holomorphic coordinate system $(z_1,\ldots,z_n)$. The same is true of the subspace spanned by $\partial/\partial \bar{z}_1,\ldots,\partial/\partial \bar{z}_n$.*

Proof. Let $(z'_1,\ldots,z'_n) = (x'_1,y'_1,\ldots,x'_n,y'_n)$ be another holomorphic coordinate system in a neighborhood of p. Then at p for any $\ell = 1,\ldots,n$

$$\frac{\partial}{\partial z_\ell} = \frac{1}{2}\left(\frac{\partial}{\partial x'_\ell} - i\frac{\partial}{\partial x'_\ell}\right) = \frac{1}{2}\left(\sum_{j=1}^n \frac{\partial x_j}{\partial x'_j}\frac{\partial}{\partial x_j} + \sum_{j=1}^n \frac{\partial y_j}{\partial x'_\ell}\frac{\partial}{\partial y_j}\right)$$

$$- \frac{i}{2}\left(\sum_{j=1}^n \frac{\partial x_j}{\partial y'_\ell}\frac{\partial}{\partial x_j} + \sum_{j=1}^n \frac{\partial y_j}{\partial y'_\ell}\frac{\partial}{\partial y_j}\right)$$

$$= \frac{1}{2}\sum_{j=1}^n\left(\frac{\partial x_j}{\partial x'_\ell} - i\frac{\partial x_j}{\partial y'_\ell}\right)\frac{\partial}{\partial x_j} - \frac{1}{2}\sum_{j=1}^n\left(\frac{\partial y_i}{\partial y'_\ell} + i\frac{\partial y_i}{\partial x'_\ell}\right)$$

$$= \sum_{j=1}^n\left(\frac{\partial x_j}{\partial x'_\ell} - i\frac{\partial x_j}{\partial y'_\ell}\right)\left[\frac{1}{2}\left(\frac{\partial}{\partial x_j} - i\frac{\partial}{\partial y_\ell}\right)\right]$$

since

$$\frac{\partial x_j}{\partial x'_\ell} - i\frac{\partial x_j}{\partial y'_\ell} = \frac{\partial y_j}{\partial y'_j} + i\frac{\partial y_j}{\partial x'_\ell}$$

by the Cauchy-Riemann equations. Thus

$$\frac{\partial}{\partial z'_\ell} = \sum_{j=1}^n\left(\frac{\partial x_j}{\partial x'_\ell} - i\frac{\partial x_j}{\partial y'_\ell}\right)\frac{\partial}{\partial z_j}$$

so $\partial/\partial z'_\ell$ belongs to the complex subspace spanned by $\partial/\partial z_1,\ldots,\partial/\partial z_n$.

The proof for the subspace spanned by $\partial/\partial\bar{z}_1,\ldots,\partial/\partial\bar{z}_n$ is similar and will be omitted.

If f is a real C^∞ function defined in a neighborhood of p, then each element of M_p^C acts on f by complex linear extension of the action of the elements of M_p on f by differention:

$$(v + iw)f = (vf) + i(wf) \quad v,w \in M_p .$$

A second complex linear extension definition gives an action of each element of M_p^C on complex valued functions defined in a neighborhood of p which are C^∞ considered as functions into R^2:

$$(v + iw)(f) = (v + iw)\mathrm{Re} f + i(v + iw)\mathrm{Im} f$$

$$= [v(\mathrm{Re} f) - w(\mathrm{Im} f)] + i[w(\mathrm{Re} f) + v(\mathrm{Im} f)]$$

for $v,w \in M_p$, Ref, Imf as usual. Note that in this sense the last equation in the proof of Lemma 2 can be rewritten

$$\frac{\partial}{\partial z'_\ell} = \sum_{j=1}^n \left(\frac{\partial}{\partial z'_\ell} z_j\right) \frac{\partial}{\partial z_j}$$

since

$$\frac{\partial}{\partial z'_\ell} z_j = \frac{1}{2}\left(\frac{\partial}{\partial x'_\ell} - i\frac{\partial}{\partial y'_\ell}\right)(x_j + iy_j)$$

$$= \frac{1}{2}\left(\frac{\partial x_j}{\partial x'_\ell} + \frac{\partial y_j}{\partial y'_\ell} - i\frac{\partial x_j}{\partial y'_\ell} + i\frac{\partial y_j}{\partial x'_\ell}\right) = \left(\frac{\partial x_j}{\partial x'_\ell} - i\frac{\partial x_j}{\partial y'_\ell}\right),$$

the last equality following from again applying the Cauchy-Riemann equations. Similar computations can be used to show that

$$\frac{\partial}{\partial \bar{z}'_\ell} = \sum_{j=1}^n \left(\frac{\partial}{\partial \bar{z}'_\ell} \bar{z}_j\right) \frac{\partial}{\partial \bar{z}_j}$$

where the $\bar{z}_j$ in the parenthetical expression denotes the function $x_j - iy_j$.

LEMMA 3. *Let f be a complex valued function defined in a neighborhood of p in M which is C^∞ as a function into R^2 and $(z_1,\ldots,z_n)$ be a holomorphic coordinate system in a neighborhood of p. Then*

(a) *for all $\ell = 1,\ldots,n$*

$$\overline{\left(\frac{\partial}{\partial z_\ell} \bar f\right)} = \frac{\partial}{\partial \bar z_\ell} f$$

(where $\overline{}$ denotes complex conjugation).

(b) *f is holomorphic on the domain of definition of the coordinate system $(z_1,\ldots,z_n)$ if and only if everywhere on the domain*

$$\frac{\partial}{\partial \bar z_\ell} f = 0 \text{ for all } \ell = 1,\ldots,n .$$

Proof. The assertion (a) follows immediately from the definitions. To prove (b), recall that f is holomorphic if and only if $f_* J_M = J_{R^2} f_*$ everywhere on the domain. Since

$$J\left(\frac{\partial}{\partial x_\ell}\right) = \frac{\partial}{\partial y_\ell} \text{ and } J\left(\frac{\partial}{\partial y_\ell}\right) = -\frac{\partial}{\partial x_\ell},$$

f is holomorphic if and only if

$$f_*\left(\frac{\partial}{\partial y_\ell}\right) = J_{R^2} f_*\left(\frac{\partial}{\partial x_\ell}\right)$$

and

$$f_*\left(-\frac{\partial}{\partial x_\ell}\right) = J_{R^2} f_*\left(\frac{\partial}{\partial y_\ell}\right).$$

Now

$$f_*\left(\frac{\partial}{\partial x_\ell}\right) = \left(\frac{\partial}{\partial x_\ell}(\mathrm{Re} f), \frac{\partial}{\partial x_\ell}(\mathrm{Im} f)\right)$$

$$f_*\left(\frac{\partial}{\partial y_\ell}\right) = \left(\frac{\partial}{\partial y_\ell}(\text{Re} f), \frac{\partial}{\partial y_\ell}(\text{Im} f)\right)$$

so

$$J_{\mathbb{R}^{2f}*}\left(\frac{\partial}{\partial x_\ell}\right) = \left(-\frac{\partial}{\partial x_\ell}(\text{Im} f), \frac{\partial}{\partial x_\ell}(\text{Re} f)\right)$$

$$J_{\mathbb{R}^{2f}*}\left(\frac{\partial}{\partial y_\ell}\right) = \left(-\frac{\partial}{\partial y_\ell}(\text{Im} f), \frac{\partial}{\partial y_\ell}(\text{Re} f)\right).$$

Thus the necessary and sufficient condition for f to be holomorphic is that

$$\frac{\partial}{\partial x_\ell}(\text{Re} f) = \frac{\partial}{\partial y_\ell}(\text{Im} f) \quad \text{and} \quad \frac{\partial}{\partial x_\ell}(\text{Im} f) = -\frac{\partial}{\partial y_\ell}(\text{Re} f).$$

Since

$$\frac{\partial}{\partial \bar{z}_\ell}f = \frac{1}{2}\left(\frac{\partial}{\partial x_\ell}(\text{Re} f) - \frac{\partial}{\partial y_\ell}(\text{Im} f)\right) + \frac{i}{2}\left(\frac{\partial}{\partial x_\ell}(\text{Im} f) + \frac{\partial}{\partial y_\ell}(\text{Re} f)\right)$$

the conclusion now follows.

The process of complexifying the tangent space and relating this complexification to holomorphic coordinate systems has an analogue in the case of the cotangent space:

DEFINITION. Let $(z_1,\ldots,z_n)$ be a holomorphic coordinate system defined in a neighborhood of a point $p \in M$. Then the elements $dz_i, d\bar{z}_i \in (M^*_p)^C$ are given by

$$dz_i \stackrel{\text{Def}}{=} dx_i + idy$$
$$d\bar{z}_i \stackrel{\text{Def}}{=} dx_i - idy.$$

Here M^*_p is the real cotangent space of M at p.

Since $dx_i = (dz_i + d\bar{z}_i)/2$ and $dy_i = (d\bar{z}_i - dz_i)/2$, it follows from Lemma 1 that $dz_1,\ldots,dz_n, d\bar{z}_1,\ldots,d\bar{z}_n$ form a complex basis for $(M^*_p)^C$.

The previous definition is a special case of the following definition:

DEFINITION. Let f be a complex-valued function defined in a neighborhood of $p \in M$ which is C^∞ considered as a function into R^2. Then $df \in (M^*_p)^C$ is given by

$$df \stackrel{\text{Def}}{=} d(\text{Re}f) + id(\text{Im}f).$$

LEMMA 4. *For any f as in the previous definition, at the point $p \in M$:*

$$df = \sum_{i=1}^n \left(\frac{\partial}{\partial z_i} f\right) dz_i + \sum_{i=1}^n \left(\frac{\partial}{\partial \bar{z}_i} f\right) d\bar{z}_i$$

for any holomorphic coordinate system $(z_1,\ldots,z_n)$ defined in a neighborhood of p.

Proof.

$$\frac{\partial}{\partial z_i} f = \frac{1}{2}\left(\frac{\partial}{\partial x_i} - i\frac{\partial}{\partial x_i}\right) f$$

$$\left(\frac{\partial}{\partial z_i} f\right) dz_i = \left(\frac{\partial}{\partial z_i} f\right)(dx_i + idy_i)$$

$$= \frac{1}{2}\left[\left(\frac{\partial}{\partial x_i} f\right) dx_i + \left(\frac{\partial}{\partial y_i} f\right) dy_i\right] + \frac{i}{2}\left[\left(\frac{\partial}{\partial x_i} f\right) dy_i - \left(\frac{\partial}{\partial y_i} f\right) dx_i\right].$$

Similarly

$$\left(\frac{\partial}{\partial \bar{z}_i} f\right) d\bar{z}_i = \frac{1}{2}\left[\left(\frac{\partial}{\partial x_i} f\right) dx_i + \left(\frac{\partial}{\partial y_i} f\right) dy_i\right]$$

$$- \frac{i}{2}\left[\left(\frac{\partial}{\partial x_i} f\right) dy_i - \left(\frac{\partial}{\partial y_i} f\right) dx_i\right].$$

Thus

$$\left(\frac{\partial}{\partial z_i} f\right) dz_i + \left(\frac{\partial}{\partial \bar{z}_i} f\right) d\bar{z}_i = \left(\frac{\partial}{\partial x_i} f\right) dx_i + \left(\frac{\partial}{\partial y_i} f\right) dy_i$$

from which the formula of the lemma follows.

LEMMA 5. *The subspace of* $(M^*_p)^C$ *spanned by* $dz_1,\ldots,dz_n$ *is independent of the holomorphic coordinate system defined in a neighborhood of* p. *The same is true of the subspace of* $(M^*_p)^C$ *spanned by* $d\bar{z}_1,\ldots,d\bar{z}_n$.

Proof. Let $(z'_1,\ldots,z'_n)$, $(z_1,\ldots,z_n)$ be two holomorphic coordinate systems each defined in a neighborhood of $p \in M$. Then by Lemma 4

$$dz'_j = \sum_{j=1}^n \left(\frac{\partial}{\partial z_j} z'_i\right) dz_j + \sum_{j=1}^n \left(\frac{\partial}{\partial \bar{z}_j} z'_i\right) d\bar{z}_j .$$

Lemma 3 implies that $(\partial/\partial \bar{z}_j) z^*_i = \overline{(\partial/\partial z_j)\bar{z}_i} = 0$, so that dz'_i is a linear combination of the dz_j's. Similarly, $d\bar{z}_i$ is a linear combination of $d\bar{z}_j$'s.

An element $\omega_1 + i\omega_2 \in (M^*_p)^C$, $\omega_1,\omega_2 \in M^*_p$, gives rise to a complex linear functional on $(M^*_p)^C$ by taking

$$(\omega_1 + i\omega_2)(v + iw) = \omega_1(v) - \omega_2(w) + i\omega_2(v) + i\omega_1(w)$$

for $v,w \in M_p$. $(M^*_p)^C$ can be thus identified with a subspace of the complex dual of M_p^C, and by dimensionality considerations this subspace is in fact the whole of the complex dual of M_p^C. Note that using this identification procedure, $dz_1,\ldots,dz_n$, $d\bar{z}_1,\ldots,d\bar{z}_n$ is the complex basis of $(M^{*p})^C$ which is dual to the basis $\partial/\partial z_1,\ldots,\partial/\partial z_n, \partial/\partial \bar{z}_1,\ldots,\partial/\partial \bar{z}_n$ of $(M_p)^C$ since

$$dz_i\left(\frac{\partial}{\partial \bar{z}_j}\right) = \frac{1}{2}(dx_i + idy_i)\left(\frac{\partial}{\partial x_j} + i\frac{\partial}{\partial y_j}\right)$$

$$= \frac{1}{2}(\delta_{ij} - \delta_{ij}) + i0 = 0$$

while

$$dz_i\left(\frac{\partial}{\partial z_j}\right) = \frac{1}{2}(dx_i + idy_i)\left(\frac{\partial}{\partial x_j} - i\frac{\partial}{\partial y_j}\right)$$

$$= \frac{1}{2}(\delta_{ij} + \delta_{ij}) + i(0) = \delta_{ij}$$

and similarly

$$d\bar{z}_i\left(\frac{\partial}{\partial z_j}\right) = 0$$

while

$$d\bar{z}_i\left(\frac{\partial}{\partial \bar{z}_j}\right) = \delta_{ij}.$$

(Here $\delta_{ij} = 0$ if $i \neq j = 1$ if $i = j$.)

The definition of $df \in (M^*_p)^C$, where f is complex valued, combined with the definition of the action of the elements of $(M^*_p)^C$ on M^C_p yields immediately that for all $V \in M^C_p$

$$df(V) = Vf$$

where Vf is defined as previously. The duality of $dz_1,\ldots,dz_n$, $d\bar{z}_1,\ldots,d\bar{z}_n$ and $\partial/\partial z_1,\ldots,\partial/\partial z_1,\ldots,\partial/\partial z_n$, $\partial/\partial \bar{z}_1,\ldots,\partial/\partial \bar{z}_n$ can be interpreted from this viewpoint also, since $(\partial/\partial z_i)\bar{z}_j = 0$, $(\partial/\partial z_i)z_j = \delta_{ij}$, etc.

5. Complex-Valued Differential Forms

DEFINITION. A *complex-valued r-form* (or *complex r-form*) on a real vector space V is an element of $(\Lambda^r V^*)^C$, where $\Lambda^r V^* = $ the real vector space of real r-forms on V. The (complex) *wedge product* $\wedge : (\Lambda^r V^*)^C \times (\Lambda^s V^*)^C \to (\Lambda^{r+s} V^*)^C$ is the complex linear extension of the real wedge product $\wedge : (\Lambda^r V^*) \times (\Lambda^s V^*) \to \Lambda^{r+s} V^*$.

The complex wedge product has the associativity and skewcommutativity properties of the real wedge product.

LEMMA 1. *If V is a real vector space of dimension N, then $(\Lambda^r V^*)^C$ has complex dimension (the binomial coefficient) $\binom{N}{r}$; and if $\{\omega_1,\ldots,\omega_N\}$ is any (complex) basis for $(V^*)^C$ then the set of forms*

$$\{\omega_{i_1} \wedge \cdots \wedge \omega_{i_r} \mid 1 \le i_1 < i_2 < \cdots < i_r \le N\}$$

is a basis for $(\Lambda^r V^*)^C$.

Proof. let $\theta_1,\ldots,\theta_N$ be a (real) basis for V^*. Then the set $\{\theta_{i_1} \wedge \cdots \wedge \theta_{i_r} \mid 1 \le i_1 < \cdots < i_r \le N\}$ is a basis for $\Lambda^r V^*$, which thus has dimension $\binom{N}{r}$. Hence $(\Lambda^r V^*)^C$ has complex dimension $\binom{N}{r}$ by Lemma 1 of §4. Since the elements $\omega_i \wedge \cdots \wedge \omega_{i_r}$, $1 \le i_1 < \cdots < i_r \le N$, are $\binom{N}{r}$ in number, they necessarily form a complex basis of $(\Lambda_r V^*)^C$ if they generate $(\Lambda_r V^*)^C$. Now, for each i, $\theta_i = \sum_{j=i}^N \alpha_{ij}\omega_j$ for some (uniquely determined) complex numbers α_{ij}, because each $\theta_i \in (V^*)^C$ and the ω_j's are a basis for $(V^*)^C$. Hence

$$\theta_{i_1} \wedge \cdots \wedge \theta_{i_r} = \left(\sum_{j=1}^N \alpha_{i_1 j}\omega_j\right) \wedge \cdots \wedge \left(\sum_{j=1}^N \alpha_{i_r j}\omega_j\right),$$

from which equation it follows that $\theta_{i_1} \wedge \cdots \wedge \theta_{i_r}$ is a linear combination with complex coefficients of wedge products of ω_j's. Since any such wedge product is $\pm$ a wedge product of the form $\omega_{j_1} \wedge \cdots \wedge \omega_{j_r}$, $1 \le j_1 < \cdots < j_r \le N$, each $\theta_{i_1} \wedge \cdots \wedge \theta_{i_r}$ belongs to the complex subspace of $(\Lambda^r V^*)^C$ generated by $\{\omega_{j_1} \wedge \cdots \wedge \omega_{j_r} \mid 1 \le j_1 < j_2 < \cdots < j_r \le N\}$. Hence this subspace is in fact all of $(\Lambda^r V^*)^C$.

If M is a complex manifold of complex dimension n and $(z_1,\ldots,z_n)$ a holomorphic coordinate system in a neighborhood of a point $p \in M$, then $\{dz_1,\ldots,dz_n, d\bar{z}_1,\ldots, d\bar{z}_n\}$ is a basis for $(M^*_p)^C$. Thus it follows from Lemma 1 that

$$\{dz_{i_1} \wedge \cdots \wedge dz_{i_p} \wedge d\bar{z}_{j_1} \wedge \cdots \wedge d\bar{z}_{j_q} \mid 1 \le i_1 < i_2 \cdots < i_p \le n,$$
$$1 \le j_i < j_2 < \cdots < j_q \le n, p + q = r\}$$

is a basis for $(\Lambda^r M^*_p)C$.

Note: We have to rely on context to distinguish the integer p from the point $p \in M$ in what follows. Twenty-six letters just are not enough! Apologies.

LEMMA 2. *If $(z_1,\ldots,z_n)$ and $(w_1,\ldots,w_n)$ are two holomorphic coordinate systems both defined in a neighborhood of $p \in M$, then $dz_{i_1} \wedge \cdots \wedge dz_{i_p} \wedge d\bar{z}_{j_1} \wedge \cdots \wedge d\bar{z}_{j_q}$ in $(\Lambda^{p+q} M^*_p)^C$ is a linear combination with complex coefficients of forms of the type $dw_{k_1} \wedge \cdots \wedge dw_{k_p} \wedge d\bar{w}_{\ell_1} \wedge \cdots \wedge d\bar{w}_{\ell_q}$ (p and q are here fixed throughout).*

Proof. Each dz_i at p is a linear combination of dw_j's and each $d\bar{z}_i$ a linear combination of $d\bar{w}_j$'s (Lemma 5, §4); the result follows directly.

DEFINITION. An element ω of $(\Lambda^r M^*_p)^C$ is said to be of *type* (p,q), $p + q = r$, if for some (and hence by Lemma 2 any) holomorphic coordinate system $(z_1,\ldots,z_n)$ defined in a neighborhood of p, Ω belongs to the subspace of $(\Lambda^r M^*_p)^C$ spanned by the set

$$\{dz_{i_1} \wedge \cdots \wedge dz_{i_p} \wedge d\bar{z}_{j_1} \wedge \cdots \wedge d\bar{z}_{j_q} \mid 1 \leq i_1 < i_2 < \cdots < i_p \leq n,$$
$$1 \leq j_1 < j_2 < \cdots < j_q \leq n\}.$$

It follows from Lemmas 1 and 2 that any element Ω of $(\Lambda^r M^*_p)^C$ can be expressed as a sum of elements $\Omega^{(p,q)}$, $p + q = r$, $p \geq 0$, $q \geq 0$, of type (p,q); moreover, the elements $\Omega^{(p,q)}$ are uniquely determined. Thus well-defined complex linear maps $\pi_{(p,q)} : (\Lambda^r M^*_p)^C \to (\Lambda^r M^*_p)C$ can be obtained by setting $\pi_{(p,q)} = \Omega^{(p,q)}$. Then for any Ω, $\Omega = \sum_{p+q=r, p\geq 0, q\geq 0} \pi_{(p,q)} \Omega$.

The elements of $(\Lambda^r M^*_p)^C$ act on r-tuples $v_1,\ldots,v_r$ of elements of M_p as complex-valued real multilinear alternating mappings. By complex multilinear extension, the elements of $(\Lambda^r M^*_p)^C$ can be taken to act on r-tuples of elements of M_p^C now as complex-valued complex multilinear alternating mappings.

A real linear conjugation operation on $(\Lambda^r M^*_p)^C$ can be defined as a special case of a general conjugation on V^C, V any real vector space $\overline{v + iw} = v - iw$, $v,w \in V$.

This operation is conjugate linear relative to the complex vector space structure on V^C. An element of V^C is an element of $V \subset V^C$ if and only if it is invariant under the conjugation operation $\overline{}$. Note that if an element of $(\Lambda^r M^*_p)^C$ is of type (p,q) then its conjugate is of type (q,p) since $\overline{dz_i} = d\bar{z}_i$ and $\overline{\Omega_1 \wedge \Omega_2} = \overline{\Omega}_1 \wedge \overline{\Omega}_2$ for any elements Ω_1, Ω_2 of $(\Lambda^r M^*_p)^C$.

DEFINITION. A (C^∞) complex-value r-form ω on M is a mapping $\omega : M \to \cup_{p \in M} (\Lambda^r M^{*p})^C$ such that

(a) $\omega(p) \in (\Lambda^r M^*_p)^C$ for all $p \in M$

(b) ω is C^∞ in the sense that when ω is expressed in terms of a

$$\{dz_{i_1} \wedge \cdots \wedge dz_{i_p} \wedge d\bar{z}_{j_1} \wedge \cdots \wedge d\bar{z}_{j_q} \mid 1 \leq i_1 < \cdots < i_p \leq n,$$
$$1 \leq j_1 < \cdots < j_q \leq n, p + q = r\}$$

basis, the coefficients are C^∞ (complex-valued) functions of $x_1, y_1, \ldots, x_n, y_n$ on the domain of holomorphic coordinate system $(z_1, \ldots, z_n)$. The complex vector space of all complex-valued r forms on M will be denoted $\Omega^r(M)$. An r-form $\omega \in \Omega^r(M)$ is of type (p,q) if $\omega(p)$ if of type (p,q) in $(\Lambda^r M^*_p)^C$ for every $p \in M$. The vector subspace of $\Omega^r(M)$ consisting of all r-forms of type (p,q) will be denoted $\Omega^{p,q}(M)$.

It is easy to verify that a complex-valued form ω which has the property that $\omega(p) \in \Lambda^r M^*_p$ for every $p \in M$ is a (real) C^∞ form in the usual sense and that, on the other hand, any real C^∞ form ω on M can be considered to be a complex-valued form. Moreover, since $dz_1, \ldots, dz_n, d\bar{z}_1, \ldots, d\bar{z}_n$ is dual to $\partial/\partial z_1, \ldots, \partial/\partial z_n, \partial/\partial \bar{z}_1, \ldots, \partial/\partial \bar{z}_1$, the condition b) in the definition of a complex-valued r-form can be reformulated as the requirement that

$$\omega(p) \left(\frac{\partial}{\partial z_{i_1}} \bigg|_p, \ldots, \frac{\partial}{\partial z_{i_p}} \bigg|_p, \frac{\partial}{\partial \bar{z}_{j_1}} \bigg|_p, \ldots, \frac{\partial}{\partial \bar{z}_{j_q}} \bigg|_p \right), p + q = r,$$

be a C^∞ (complex-valued) function of $p \in M$ on the domain of $(z_1, \ldots, z_n)$ for any set of indices $i_1,\ldots,i_p, j_1,\ldots,j_q$. It follows that if $v_1,\ldots,v_r$ are any r complex linear combinations of $\partial/\partial z_1,\ldots,\partial/\partial z_n, \partial/\partial \bar{z}_1,\ldots,\partial/\partial \bar{z}_n$ with C^∞ coefficients then $\omega(p)(v_1|_p,\ldots,v_r|_p)$ is a C^∞ function of p. In particular,

$$\omega(p)\left(\left.\frac{\partial}{\partial x_1}\right|_p, \left.\frac{\partial}{\partial y_1}\right|_p,\ldots,\left.\frac{\partial}{\partial x_n}\right|_p, \left.\frac{\partial}{\partial y_n}\right|_p\right)$$

is a C^∞ function. An extended (to arbitrary C^∞ real coordinate systems) converse of this last result also holds: $\omega : M \to \cup_{p \in M} (\Lambda^r M^*_p)^C$ with $\omega(p) \in (\Lambda^r M^*_p)C$ is C^∞ if and only if it is C^∞ in real C^∞ coordinate expressions.

By complex linearity, the operator d can be extended to $\Omega^r(M)$: Explicitly, let $(u_1,\ldots,u_{2n})$ be a real coordinate system in a neighborhood of $p \in M$ and write (uniquely)

$$\omega = \sum_{1 \leq i_1 < \cdots < i_r \leq 2n} f_{i_1,\ldots,i_r}\, du_{i_1} \wedge \cdots \wedge du_{i_r},$$

where the f's are C^∞ complex-valued. Then

$$d\omega = \sum_{1 \leq i_1 < \cdots < i_r \leq 2n} df_{i_1,\ldots,i_r} \wedge du_{i_1} \wedge \cdots \wedge du_{i_r}$$

where $df_{i_1,\ldots,i_r}$ is defined as in §4. The usual argument from the real case shows that $d\omega \in \Omega^{r+1}(M)$ thus defined is independent of the choice of the real coordinate system $(u_1,\ldots,u_{2n})$. Also from the real case it follows easily that d thus extended to complex-valued forms satisfies

$$d(\omega_1 \wedge \omega_2) = d\omega_1 \wedge \omega_2 + (-1)^r \omega_1 \wedge d\omega_2$$

for $\omega \in \Omega^r(M)$.

LEMMA 3.

$$d(f\, dz_{i_1} \wedge \cdots \wedge dz_{i_p} \wedge d\bar{z}_{j_1} \wedge \cdots \wedge d\bar{z}_{j_q})$$

$$= \sum_{i=1}^n \left(\frac{\partial f}{\partial z_i}\, dz_i \wedge dz_{i_1} \wedge \cdots \wedge dz_{i_p} \wedge d\bar{z}_{j_1} \wedge \cdots \wedge d\bar{z}_{j_q}\right)$$

$$+ \sum_{i=1}^{n} \left((-1)^p \frac{\partial f}{\partial \bar{z}_i} dz_{i_1} \wedge \cdots \wedge dz_{i_p} \wedge d\bar{z}_i \wedge d\bar{z}_{j_1} \wedge \cdots \wedge d\bar{z}_{j_q} \right).$$

Proof. Since $d(dz_\ell) = d(d\bar{z}_\ell) = 0$ for all $\ell = 1,\ldots,n$, the formula given follows by repeated application of $d(\omega_1 \wedge \omega_2) = d\omega_1 \wedge \omega_2 + (-1)^r \omega_1 \wedge d\omega_2$, $\omega_1, \omega_2 \in \Omega^r$, together with Lemma 4 of §4.

Lemma 3 implies immediately that, for $\omega \in \Omega^{p,q}(M)$,

$$d\omega \in \Omega^{p+1,q}(M) + \Omega^{p,q+1}(M).$$

DEFINITION. $\partial : \Omega^{p,q}(M) \to \Omega^{p+1,q}(M)$ is by definition $\pi_{p+1,q} \circ d : \Omega^{p,q}(M) \to \Omega^{p+1,q}(M)$. $\bar{\partial} : \Omega^{p,q}(M) \to \Omega^{p,q+1}(M)$ is $\pi_{p,q+1} \circ d : \Omega^{p,q}(M) \to \Omega^{p,q+1}(M)$. More generally,

$$\partial : \Omega^r(M) \to \Omega^{r+1}(M) \text{ is the operator } \sum_{p+q=r} \pi_{p+1,q}\, d\, \pi_{p,q}$$

and

$$\bar{\partial} : \Omega^r(M) \to \Omega^{r+1}(M) \text{ is } \sum_{p+q=r} \pi_{p,q+1}\, d\, \pi_{p,q}.$$

In case the meaning is clear, the symbols ∂ and $\bar{\partial}$ will be used without explicit notation of their domain, as is usually done with the operator d.

LEMMA 4. $d = \partial + \bar{\partial}$, $\partial^2 = \bar{\partial}^2 = \partial\bar{\partial} + \bar{\partial}\partial = 0$. If $\omega \in \Omega^{p,q}(M)$ then $\bar{\omega} \in \Omega^{q,p}(M)$ and $\bar{\partial}\bar{\omega} = \overline{\partial \omega}$.

Proof. All the assertions follow easily from the definitions of ∂ and $\bar{\partial}$, the fact that $d^2 = 0$, the direct sum decomposition of $\Omega^r(M) = \sum_{p+q=r} \Omega^{p,q}(M)$, and the behavior of types under conjugation. The details are left to the reader.

Lemma 3 provides a computational description of ∂ and $\bar{\partial}$:

$$\partial(f\, dz_{i_1} \wedge \cdots \wedge dz_{i_p} \wedge d\bar{z}_{j_1} \wedge \cdots \wedge d\bar{z}_{j_q})$$

$$= \sum_{i=1}^{n} \frac{\partial f}{\partial z_i} dz_i \wedge dz_{i_1} \wedge \cdots \wedge dz_{i_p} \wedge d\bar{z}_{j_1} \wedge \cdots \wedge d\bar{z}_{j_q}$$

$$\bar{\partial}(f\, dz_{i_1} \wedge \cdots \wedge dz_{i_p} \wedge d\bar{z}_{j_1} \wedge \cdots \wedge d\bar{z}_{j_q})$$

$$= \sum_{i=1}^{n} (-1)^p \frac{\partial f}{\partial \bar{z}_i}\, dz_{i_1} \wedge \cdots \wedge dz_{i_p} \wedge d\bar{z}_i \wedge d\bar{z}_{j_1} \wedge \cdots \wedge d\bar{z}_{j_q}\,.$$

These formulae can be used to provide computational proof of Lemma 4.

6. Hermitian and Kähler Metrics in Complex Notation

The subspace of M_p^C spanned by $\partial/\partial z_1|_p, \ldots, \partial/\partial z_n|_p$ (where $(z_1,\ldots,z_n)$ is a holomorphic coordinate system in a neighborhood of $p \in M$) is called the *holomorphic tangent space* at p. *It is a complex vector space of dimension* n. By Lemma 2, §4, it is well defined independently of the choice of $(z_1,\ldots,z_n)$ coordinates. Notation: M_p^h.

If g is a Hermitian metric on M (i.e., a J-invariant Riemannian metric), then g determines a Hermitian metric on each holomorphic tangent space M_p^h in the conventional sense of complex linear algebra as follows: First note that g extends by complex linearity to be a complex bilinear form $\hat{g}$ on M_p^C. (This form cannot be positive definite!) Define for $V, W \in M_p^h$

$$g(V,W) = \hat{g}(V,\overline{W})\,.$$

Then g on M_p^h so defined is a positive definite Hermitian metric on M_p^h. (Note: The use— for both M_p and M_p^h— of the same symbol g can lead to no confusion, especially since, for real vectors, the conjugation in the definition of g on M_p^h would be of no effect.)

Define, given $(z_1,\ldots,z_n)$ holomorphic coordinates,

$$g_{i\bar{j}} = g\left(\frac{\partial}{\partial z_i}, \frac{\partial}{\partial z_j}\right) = \hat{g}\left(\frac{\partial}{\partial z_i}, \overline{\frac{\partial}{\partial z_j}}\right)\,.$$

Then

$$g_{i\bar{j}} = \frac{1}{4}\hat{g}\left(\frac{\partial}{\partial x_i} - \sqrt{-1}\,\frac{\partial}{\partial y_i}, \frac{\partial}{\partial x_j} + \sqrt{-1}\,\frac{\partial}{\partial y_j}\right)$$

$$= \frac{1}{4} \left[g\left(\frac{\partial}{\partial x_i}, \frac{\partial}{\partial x_j} \right) + g\left(\frac{\partial}{\partial y_i}, \frac{\partial}{\partial y_j} \right) \right] + \frac{\sqrt{-1}}{4} \left[g\left(\frac{\partial}{\partial x_i}, \frac{\partial}{\partial y_j} \right) - g\left(\frac{\partial}{\partial y_i}, \frac{\partial}{\partial x_j} \right) \right]$$

$$= \frac{1}{2} g\left(\frac{\partial}{\partial x_i}, \frac{\partial}{\partial x_j} \right) + \frac{\sqrt{-1}}{2} g\left(\frac{\partial}{\partial x_i}, \frac{\partial}{\partial y_j} \right)$$

because

$$g\left(\frac{\partial}{\partial y_i}, \frac{\partial}{\partial y_j} \right) = g\left(J \frac{\partial}{\partial x_i}, J \frac{\partial}{\partial x_j} \right) = g\left(\frac{\partial}{\partial x_i}, \frac{\partial}{\partial x_j} \right)$$

and

$$g\left(\frac{\partial}{\partial y_i}, \frac{\partial}{\partial x_j} \right) = g\left(J \frac{\partial}{\partial x_i}, \frac{\partial}{\partial x_j} \right) = - g\left(\frac{\partial}{\partial x_i}, J \frac{\partial}{\partial x_j} \right) = - g\left(\frac{\partial}{\partial x_i}, \frac{\partial}{\partial y_i} \right).$$

These formulae express g on M_p^h in terms of g on M_p. They also show (when run the other way) the following: Every Hermitian metric on M_p^h arises from exactly one J-invariant (Riemannian) metric on M_p.

Of course, since all these considerations happen at one point p at a time, they are really facts about complex linear algebra!

As a tensor, the metric on $M_{\mathbb{C}}^h$ is given by

$$G = \sum_{i,j} g_{i\bar{j}} \, dz_i \otimes d\bar{z}_j$$

in the sense that*

$$g(V,W) = G(V,\overline{W}) \quad V,W \in M_p^h .$$

Associated to the tensor G ($= \hat{g}$) is a (real) 2-form

$$- \sqrt{-1} \sum_{i,j} g_{i\bar{j}} \, dz_i \wedge d\bar{z}_j .$$

---

*No complex tensor could operate on V,W to give $g(V,W)$: there must be a complex conjugation applied to W first since $g(V, + \sqrt{-1} \, W) = - \sqrt{-1} \, g(V,W)$!

Exercise. Check that this association is well defined independently of coordinate choice. This form is just the Kähler form as defined on page 3.2 (page 2 of §3).

Philosophical remark. This is one of many places where it is possible to get $\pm$ signs wrong, or to gain or lose factors of 1/2 or 1/4. Usually it does not matter. But in cases where it does (later: positive vs. negative curvature), it is safest to check all computations by simple examples. There are many papers that are supposed to be about positive bundle (definition later) that are actually about negative ones, etc.

The metric is Kähler if and only if

$$0 = d\left(\sum_{i,j} g_{i\bar{j}} dz_i \wedge d\bar{z}_j\right) \Leftrightarrow 0 = \sum_{i,j,\ell} \left(\frac{\partial g_{i\bar{j}}}{\partial z_\ell} dz_\ell + \frac{\partial g_{i\bar{j}}}{\partial \bar{z}_\ell} d\bar{z}_\ell\right) dz_i \wedge d\bar{z}_j .$$

Since the terms from the dz_ℓ have type (2,1) while those from the $d\bar{z}_\ell$ have type (1,2) each sum

$$\sum \frac{\partial g_{i\bar{j}}}{\partial z_\ell} dz_\ell \wedge dz_i \wedge d\bar{z}_j \text{ and } \sum \frac{\partial g_{i\bar{j}}}{\partial \bar{z}_\ell} d\bar{z}_\ell \wedge dz_i \wedge d\bar{z}_j$$

must vanish separately if (and only if) g is a Kähler metric. It follows that: g is a Kähler metric if and only if

$$\frac{\partial g_{i\bar{j}}}{\partial z_\ell} = \frac{\partial g_{\ell\bar{j}}}{\partial z_i} \text{ and } \frac{\partial g_{i\bar{j}}}{\partial \bar{z}_\ell} = \frac{\partial g_{i\bar{\ell}}}{\partial \bar{z}_j}$$

for all i,j,ℓ. We shall use this soon, in relation to holomorphic normal coordinates.

7. Holomorphic Normal Coordinates

In this section, we want to prove the following computationally useful proposition:

PROPOSITION. *Let M be a complex manifold with Hermitian metric g, and let ω be the Kähler form of g. Let $p \in M$. Then $d\omega|_p = 0$ if and only if $\exists$ a holomorphic coordinate system $(z_1,...,z_n)$ defined in a neighborhood of p with*

(1) $$g\left(\frac{\partial}{\partial z_i}, \frac{\partial}{\partial z_j}\right)\bigg|_p = 0 \quad i \neq j$$
$$= 1 \quad i = j$$

and

(2) $$d\left[g\left(\frac{\partial}{\partial z_i}, \frac{\partial}{\partial z_j}\right)\right]\bigg|_p = 0 \text{ all } i,j.$$

Proof. Conditions [(1) and (2)] imply that $d\omega|_p = 0$ by the formulae of §6. The converse is more complicated. We do it as follows:

Choose coordinates (holomorphic) $w_1,\ldots,w_n$ with $p \sim (0,\ldots,0)$. Changing the w coordinates by a constant coefficient linear transformation, we can assume condition (1) to hold. To arrange for condition (2) to hold, we introduce $z_1,\ldots,z_n$ defined by

$$z_i + \sum_{j,k} \alpha^i_{jk} z_j z_k = w_i,$$

where α^i_{jk} are complex constants to be chosen later and which are to satisfy $\alpha^i_{jk} = \alpha^i_{kj}$. Note that for each choice of α's, the equations define a holomorphic coordinate system $(z_1,\ldots,z_n)$ in some (smaller) neighborhood of p: this follows from the holomorphic implicit function theorem (actually, inverse function theorem). By the (holomorphic) chain rule:

$$\frac{\partial}{\partial z_i} = \sum_\ell \frac{\partial w_\ell}{\partial z_i} \frac{\partial}{\partial w_\ell} = \sum_\ell \left[\delta_{i\ell} + \sum_j (\alpha^\ell_{ij} + \alpha^\ell_{ji}) z_j\right] \frac{\partial}{\partial w_\ell}$$

and

$$\frac{\partial}{\partial \bar{z}_i} = \sum_\ell \left[\delta_{ij} + \sum_j (\bar{\alpha}^\ell_{ij} + \bar{\alpha}^\ell_{ji}) \bar{z}_j\right] \frac{\partial}{\partial \bar{w}_\ell}.$$

Hence, $g_{i\bar{j}}$ in z coordinates, $h_{i\bar{j}}$ for w coordinates,

$$g_{i\bar{j}} = \hat{g}\left(\sum_\ell \left[\delta_{i\ell} + 2\sum_q \alpha^\ell_{iq} z_q\right] \frac{\partial}{\partial w_\ell}, \sum_k \left[\delta_{jk} + 2\sum_m \bar{\alpha}^k_{jm} \bar{z}_m\right] \frac{\partial}{\partial \bar{w}_k}\right)$$

$$= h_{i\bar{j}} + 2\hat{g}\left(\sum_q \alpha_{iq}^\ell z_q \frac{\partial}{\partial w_\ell}, \frac{\partial}{\partial \overline{w}_j}\right) + 2\hat{g}\left(\frac{\partial}{\partial w_i}, \sum_k \overline{\alpha}_{jm}^k \overline{z}_m \frac{\partial}{\partial \overline{w}_k}\right)$$

$+$ terms quadratic in z's.

At p, $g_{i\bar{j}} = h_{i\bar{j}}$ so condition (1) remains satisfied. For condition (2), we need

$$\left.\frac{\partial g_{i\bar{j}}}{\partial z_r}\right|_p = 0 \quad \text{and} \quad \left.\frac{\partial g_{i\bar{j}}}{\partial \overline{z}_r}\right|_p = 0.$$

Again at p (where z's $=$ w's $= 0$)

(*) $$\frac{\partial g_{i\bar{j}}}{\partial z_r} = \frac{\partial h_{i\bar{j}}}{\partial z_r} + 2\overline{\alpha}_{ir}^j$$

because $\hat{g}(\partial/\partial w_i, \partial/\partial \overline{w}_j) = \delta_{ij}$ at p. Also at p

(**) $$\frac{\partial g_{i\bar{j}}}{\partial \overline{z}_r} = \frac{\partial h_{i\bar{j}}}{\partial \overline{z}_r} = \frac{\partial h_{i\bar{j}}}{\partial \overline{z}_r} + 2\overline{\alpha}_{jr}^i.$$

So from (*) and (**) condition (2) holds if

(†) $$\alpha_{ir}^j = -\frac{1}{2}\frac{\partial h_{i\bar{j}}}{\partial z_r}$$

and

(††) $$\overline{\alpha}_{jr}^i = -\frac{1}{2}\frac{\partial h_{i\bar{j}}}{\partial \overline{z}_r}.$$

If (†) holds (for all i, j, r) then so does (††) because

$$\overline{\frac{\partial h_{i\bar{j}}}{\partial \overline{z}_r}} = \frac{\partial \overline{h_{i\bar{j}}}}{\partial z_r} = \frac{\partial h_{j\bar{i}}}{\partial z_r}.$$

So we need only arrange for (†) to hold for all i, j, r at once. This will be possible if and only if the righthand side of (†) is symmetric in i and r. (Remember: We assumed α_{ir}^j was symmetric in i and r, without loss of generality.) In other words, we must have

(†††) $$\frac{\partial h_{i\bar{j}}}{\partial z_r} = \frac{\partial h_{r\bar{j}}}{\partial z_i}$$

and, if we do have this, then we can solve (†). But (†††) is true if g is Kähler (end of §6). So the proof is complete. ∎

While this proof is clear computationally, it is worthwhile to look at the situation philosophically, too. In general, Riemann normal coordinates (i.e., $\exp^{-1}$) will not be real and imaginary part of a holomorphic coordinate system, i.e., J on M_p will not be taken, in general, to J on M by $\exp_p$. Of course, since $d(\exp_p)|_p =$ identity, J at p is taken to itself. Also, if the metric is Kähler so that $DJ|_p = 0$, then [because Riemann normal coordinates are parallel-at-p (i.e. $D(\text{coord field})|_p = 0$] the exponential map is J-preserving at p to one higher order than just the 0-order, $d\exp_p|_p =$ identity. So it makes sense that there is some holomorphic coordinate system with real and imaginary parts $x_1, y_1, \ldots, x_n, y_n$ matching Riemann normal coordinates to an extra order, i.e., having covariant derivative 0 at p. This is just what condition (2) of the Proposition gives.

Exercise. Show that with $(z_1, \ldots, z_n)$ as in the Proposition and $z_j = x_j + \sqrt{-1}\, y_j$, it holds that

$$D_{\partial/\partial x_j} \frac{\partial}{\partial y_\ell} = 0 \quad D_{\partial/\partial x_j} \frac{\partial}{\partial y_\ell} = 0$$

$$D_{\partial/\partial x_j} \frac{\partial}{\partial x_\ell} = 0 \quad D_{\partial/\partial y_j} \frac{\partial}{\partial x_\ell} = 0$$

all at p, all j, ℓ.

8. Basic Examples of Kähler Manifolds

(1) $\mathbf{C}^n$ with its standard metric (as $\mathbf{R}^{2n}$). This is easily seen to be Kähler because in standard coordinates its metric coefficients are constant.

(2) $P_n\mathbf{C}$: Denote by $P : S^{2n+1} \to \mathbf{C}P^n$ the projection (to equivalence classes) of $S^{2n+1} = \{(z_1,\ldots,z_{n+1}) \in \mathbf{C}^{n+1} \mid \Sigma|z_j|^2 = 1\}$ onto $\mathbf{C}P^n$. If

$$(z_1,\ldots,z_{n+1}) = (x_1,y_1,\ldots,x_{n+1},y_{n+1}),$$

then $(x_1,\ldots,y_{n+1})$ and $(-y_1,x_1,\ldots,-y_{n+1},x_{n+1})$ together span a J-invariant subspace S of $(\mathbf{C}^{n+1})_{(x_1\cdots y_{n+1})}$ = the real tangent space of $\mathbf{C}^{n+1} = \mathbf{R}^{2n+2}$ at $(x_1,\ldots,y_{n+1})$. The orthogonal complement of this subspace S projects J-equivariantly under dP one-to-one onto the tangent space of $\mathbf{C}P^n$ at $P(z_1,\ldots,z_{n+1})$. To see this note that the two have the same real dimension $2n$. Also the vector $(-y_1,x_1,\ldots,-y_{n+1},x_{n+1})$ generates the (one dimensional) tangent space to the fibre of P through $(x_1,\ldots,y_{n+1})$ because it is the tangent to the curve

$$t \to (\cos t + i\sin t) \cdot (z_1,\ldots,z_{n+1})$$

which is the fibre. So $dP|$ orthogonal complement S is injective.

We now define a metric on $\mathbf{C}P^n$ by declaring $dP|$ orthogonal complement of S to be isometric. It is algebra to check that the resulting metric is C^∞ Hermitian. To see that it is Kähler is more tedious if one attempts direct calculation. But the situation can be simplified as follows:

First note that the unitary group acting on $\mathbf{C}^{n+1}$ induces an action $\mathbf{C}P^n$; and, as is easy to see, this action is isometric relative to the metric we have defined. It is also biholomorphic (i.e., each induced mapping is a biholomorphic map of $\mathbf{C}P^n$ to itself). Thus we need only check the vanishing of DJ (or $d\omega$) at one point of $\mathbf{C}P^n$ — because the group action is transitive and DJ (or $d\omega$) is an invariant under biholomorphic isometries. So let us check what happens at $[(1,0,\ldots,0)]$. Writing for convenience $(z_0,\ldots,z_n)$ as coordinates on $\mathbf{C}^{n+1}$ we have the coordinates on $\mathbf{C}P^n$

$$(z_1,\ldots,z_n) \to [(1,z_1,\ldots,z_n)]$$

as a holomorphic coordinate system near $[(1,0,\ldots,0)]$. If this coordinate system has $dg_{i\bar{j}}|_{(0,\ldots,0)} = 0$, then the formulae in §6 show that $DJ = 0$ (or $d\omega = 0$) at $(0,\ldots,0)$.

Now computing the length of $\partial/\partial z_i, i \geq 1$, (or the inner product $\hat{g}(\partial/\partial z_i, \partial/\partial \bar{z}_j)$) will involve computing the dP preimage in the orthogonal complement of

$$\left(\frac{1}{\sqrt{1 + z_1^2 + \cdots z_n^2}}, \ldots, \frac{z_n}{\sqrt{1 + \cdots z_n^2}} \right)$$

and J of that vector (considered as a real vector). More precisely

$$g\left(\frac{\partial}{\partial z_i}, \frac{\partial}{\partial z_i} \right) = H\left(Q\left(\frac{\partial}{\partial z_i} \right), Q\left(\frac{\partial}{\partial z_i} \right) \right)$$

where H = the Euclidean metric on $\mathbf{C}^{n+1}$ and Q = projection on the orthogonal complement of $\{(1/\sqrt{}, \ldots, z_n/\sqrt{})$ and J of that vector$\}$. A straightforward calculation involving order of magnitude argument only* shows that

$$g\left(\frac{\partial}{\partial z_i}, \frac{\partial}{\partial z_i} \right)$$

is constant to second order (i.e., has 0 first derivatives), and similarly for

$$g\left(\frac{\partial}{\partial z_i}, \frac{\partial}{\partial z_j} \right).$$

(Note: This argument also incidentally shows in detail that the metric on $\mathbf{C}P^n$ is J-invariant).

(3) B^n and its Poincare metric $B^n = \{(z_1, \ldots, z_n) \mid \Sigma |z_j|^2 < 1\})$.

There is a large group of biholomorphic mappings acting on B^n. First of all, the unitary group acts fixing the origin. It is well known (and fairly easy to see) that these are the only biholomorphic mappings that fix the origin. In addition, the maps (for each $a, a \in \mathbf{C}, |a| < 1$)

$$(z_1, \ldots, z_n) \to \left(\frac{z_1 - a}{1 - \bar{a}z_1}, \frac{z_2(1 - |a|^2)^{1/2}}{1 - \bar{a}z_1}, \ldots, \frac{z_n(1 - |a|^2)^{1/2}}{1 - \bar{a}z_1} \right)$$

*$Q(\partial/\partial z_i) = \partial/\partial z_i$ - term vanishing to 1st order that is perpendicular to $(1/\sqrt{}, \ldots, z_n/\sqrt{})$ and J of that.

are easily checked to be biholomorphic. Combining these two types of maps, we see that biholomorphic maps act transitively on B^n. We define a metric on B^n by setting the metric at $p \in B^n = \alpha^*_p g_0$, where g_0 = Euclidean metric at origin and $\alpha_p : B^n \to B^n$ is a biholomorphic map taking p to the origin. (Any two such differ by a unitary group element, so the metric is well defined.)

The metric we have defined is obviously J-invariant. (Exercise: Show it is C^∞). To see that it is Kähler, it is enough to see that $d\omega = 0$ (or $DJ = 0$) at $(0,\ldots,0)$ since all points are biholomorphically isometrically equivalent to $(0,\ldots,0)$. Also, since the unitary group acts transitively on directions at $(0,\ldots,0)$, it is enough to check that $D_X J = 0$ where $X = \partial/\partial x_1|_{(0,\ldots,0)}$. For this, consider the (standard) holomorphic coordinate system $(z_1,\ldots,z_n)$. If

$$0 = D_X \frac{\partial}{\partial x_i} = D_X \frac{\partial}{\partial y_j} \quad \text{all } i, j$$

then $D_X J = 0$ since $J(\partial/\partial x_i) = \partial/\partial y_i$.

But straightforward calculation shows that $g(\partial/\partial z_i, \partial/\partial \bar{z}_j)$ is 2nd order constant at $(0,\ldots,0)$ so $D\, \partial/\partial x_1 = D\, \partial/\partial y_i = 0$ at $(0,\ldots,0)$. ∎

9. Curvature Properties of Kähler Manifolds

Recall the Riemann curvature 3-tensor

$$R(X,Y)Z \stackrel{\text{def}}{=} (D_X D_Y - D_Y D_X - D_{[X,Y]})Z$$

and 4-tensor

$$R(X,Y,Z,W) \stackrel{\text{def}}{=} -g(R(X,Y)Z,W).$$

On a Kähler manifold, these tensors have special properties relative to J: Since J is parallel, it follows that

$$R(X,Y)(JZ) = JR(X,Y)Z.$$

From this follows

$$-R(X,Y,JZ,JW) = g(R(X,Y)JZ,JW) = g(JR(X,Y)Z,JW)$$

$$= g(R(X,Y)Z,W) = -R(X,Y,Z,W).$$

Also since $R(X,Y,Z,W) = R(Z,W,X,Y) = R(Z,W,X,Y)$ on any Riemannian manifold, we have

$$R(JX,JY,Z,W) = R(Z,W,JX,JY) = R(Z,W,X,Y) = R(X,Y,Z,W).$$

So $R(,,,)$ is J-invariant in the first two slots and in the last two.

Let X be a unit vector. The number $R(X,JX,X,JX)$ is called the *holomorphic sectional curvature* of the 2-plane spanned by X, JX. If Y is any other unit vector in this 2-plane, then $R(X,JX,X,JX) = R(Y,JY,Y,JY)$ (Exercise: Check this.) Thus holomorphic sectional curvature is a well-defined function on the J-invariant 2-planes in M_p.

The holomorphic sectional curvatures at p determine the whole curvature tensor at p. In fact, a purely algebraic kind of determination holds:

Proposition: Suppose R and T are 4-tensors on a vector space V with endomorphism $J: V \to V \ni J^2 = -1$. Suppose R and T have the symmetries of a Riemann curvature tensor (antisymmetric in 1st two, last two; symmetric under $1234 \to 3412$; 1st Bianchi) and that both are J invariant in first two and in last two shots. Suppose also that for all $X \in V$

$$R(X,JX,X,JX) = T(X,JX,X,JX).$$

Then

$$R = T.$$

Proof: It suffices to consider the case $T = 0$. Look at the map

$$(X,Y,U,V) \xrightarrow{\hat{R}} R(X,JY,U,JV) + R(X,JU,Y,JV) + R(X,JV,Y,JU).$$

This is symmetric in X, Y, U and V (check this). Also it $= 0$ for $X = Y = U = V$ since $T = R = 0$ for X, JX, X, JX etc. By polarization, $\hat{R} = 0$. With $X = U$, $Y = V$, one gets

(*) $$2R(X,JY,X,JY) + R(X,JX,Y,JY) = 0.$$

From the Bianchi identity

$$R(X,JX,Y,JY) + R(X,Y,JY,JX) + R(X,JY,JX,Y) = 0$$

so

$$R(X,JX,Y,JY) - R(X,Y,X,Y) - R(X,JY,X,JY) = 0.$$

Adding minus this to (*) gives

$$3R(X,JY,X,JY) + R(X,Y,X,Y) = 0$$

Replace Y by JY:

$$3R(X,Y,X,Y) + R(X,JY,X,JY) = 0.$$

These last two imply $R(X,Y,X,Y) = 0$. Since sec. curv $= 0 \Rightarrow R = 0$ (in a purely algebraic way), it follows that

$$R = 0. \quad \blacksquare$$

The proposition we have just proved shows that there could be at most one tensor R (with all the symmetries and J invariances) $\ni R(X,JX,X,JX) = +1$ for $X \ni \|X\| = 1$. We can in fact exhibit one such. (Note: We shall do so by just writing one down. But in principle we could *compute* one, by noting that CP^n has constant holomorphic sectional curvature, since the biholomorphic isometries act transitively on the J-invariant 2-planes. So we could just compute the curvature tensor of CP^n, and even at one point, in fact. Then we would have at most to multiply it by a constant). With g as the real J-invariant inner product set

$$R_0(X,Y,U,V) = \frac{1}{4}\{g(X,U)g(Y,V) - g(X,V)g(Y,U)$$
$$+ g(X,JU)g(Y,JV) - g(X,JV)g(Y,JU)$$

$$+ 2g(X,JY)g(U,JV)\}.$$

A tedious but easy check shows R_0 has the required symmetries and satisfies

$$R_0(X,JX,X,JX) = g(X,X)^2.$$

Also,

$$R_0(X,Y,X,Y) = \frac{1}{4}\{g(X,X)g(Y,Y) - g(X,Y)^2 + 3g(X,JY)^2\}.$$

This last formula has a pleasant geometric meaning.

Suppose P is a 2-plane. Then we define the angle α_P between P and JP by α_p = minimum angle between $X \in P$ and $Z \in JP$. Then

$$\cos(\alpha(p)) = |g(X,JY)|$$

where X, Y is an orthonormal basis for P. (Exercise: Prove this.) Thus the sectional curvature (for R_0 as curvature tensor)

$$K(P) = \frac{1}{4}(1 + 3\cos^2(\alpha_p)).$$

We can summarize all this as follows: If at a point $p \in M$, M a Kähler manifold, all J-invariant 2-planes have the same sectional curvature c then the curvature tensor at p is cR_0 and for any 2-plane P in M_p,

$$K(P) = (1/4)(1 + 3\cos^2(\alpha_1))c.$$

The next result shows that a Kähler manifold of C-dimension ≥ 2 can satisfy the previous condition at each point only in a special way:

PROPOSITION. *If M is a (connected) Kähler manifold with* $\dim_{\mathbb{R}} M = 2 \dim_{\mathbb{C}} 2M \geq 4$ *and if, at each point* $p \in M$, $R = c_p R_0$ *for some constant* c, *then in fact c is independent of p!*

Proof. Let Ric denote the Ricci tensor of M. Compute for an arbitrary $X, X \in M_p$, $\|X\| = 1$

$$\text{Ric}(X,X) = R(X,JX,X,JX) + \sum_{j=2}^{n} [R(X,Y_j,X,Y_j) + R(X,JY_j,X,JY_j)]$$

where $X, JX, Y_2, JY_2, \ldots, Y_n, JY_n$ is an orthonormal basis of M_p. Now $R(X,JX,X,JX) = c_p$ while

$$R(X,Y_j,X,Y_j) = R(X,JY_j,X,JY_j) = 1/4 \, c_p \,.$$

So

$$\text{Ric}(X,X) = c_p + 2[(n-1)/4] c_p = \left(\frac{n+1}{2}\right) c_p \,.$$

In particular, Ric(X,X) is independent of $X \in M_p$, $\|X\| = 1$. So M is Einstein (at p). By a classical theorem the quotient $c_p = \{1/(n + 1/2)\}$ [Ric/g] is independent of p. ∎

Of course, none of this discussion about independence of point has any relevance in case $n = 1$ (a 2-dimensional manifold in the sense of real dimension).

Our three basic Kähler manifolds — $\mathbf{C}^n$, $\mathbf{C}P^n$, B^n — all have constant holomorphic sectional curvature. $\mathbf{C}^n$ is of course flat: $R = 0$. $\mathbf{C}P^n$ has constant holomorphic sectional curvature (because the biholomorphic isometries act transitively on the J-invariant 2-planes). $\mathbf{C}P^n$ is simply connected and compact so it cannot admit a (necessarily complete) Riemannian metric with sectional curvature ≤ 0. Hence its holomorphic sectional curvature (which is constant) is > 0. Similarly, the complete metric on B^n cannot have positive curvature, since it would be positive bounded away from zero, contradicting the noncompactness of B^n. So $R_{B^n} = cR_0$, $c \leq 0$. It cannot be the case that $c = 0$ because then B^n would be biholomorphically isometric to $\mathbf{C}^n$: namely, the exponential map would be such a biholomorphic isometry. So $c < 0$ in this case.

These three examples in fact are all the possibilities for constant holomorphic sectional curvature, up to coverings and constant factors. Specifically, the following result holds:

PROPOSITION. *Any two complete n-dimensional* C^∞ *simply connected Kähler manifolds with the same constant positive holomorphic sectional curvature are biholomorphically isometric to each other.*

Proof. Let M_1 and M_2 be two such manifolds. Choose $p_1 \in M_1$ and $p_2 \in M_2$ and let $T: M_{p_1} \to M_{p_2}$ be an isometric linear transformation that commutes with J. (Such a transformation can be obtained by choosing orthonormal bases $X_1, JX_1, \ldots, X_n, JX_n$ and $Y_1, JY_1, \ldots, Y_n, JY_n$ and defining $T(X_j) = Y_j$, $T(JX_j) = JY_j$ for all $j = 1, \ldots, n$.) The transformation T also takes R_{p_1} to R_{p_2}, by our previous results. Moreover, since J_{M_1} and J_{M_2} are both parallel and since the curvature of M_1 (and M_2) is determined by the metric, J and the equal constant holomorphic sectional curvatures, it follows that both M_1 and M_2 have parallel curvature tensors. By simple connectivity of M_1, and standard considerations*, T must be the differential of a locally isometric surjective covering map $T: M_1 \to M_2$. T must be holomorphic (Exercise: Prove this by using parallelism of J to show $dT \circ J_{M_1} = J_{M_2} \circ dT$ at each point of M_1, since the equation holds at p). Since M_2 is simply connected, T must be injective. ■

10. Kähler Submanifolds

Let N be a complex submanifold of a complex manifold M. Since by definition $J_N|_q = $ the restriction to N_q of $J_M|_q$ (N_q being J-invariant), it follows that the restriction to N of a Hermitian metric on M is a Hermitian metric on N. Also, the Kähler form of the metric on N is the restriction to N of the Kähler form of the

---
*of Riemannian geometry.

Hermitian metric on M. Since $d_N = d_M|N$, it follows that if the metric on M is Kähler then so is the metric on N. The relationship between the curvature of the metric on M and that of N is much closer than in the Riemannian case, where, in the case of large codimension, at least, there is no relationship. The relationship in the Kähler situation comes from special properties of the second fundamental form.

Let $S(X,Y) = D_X^M Y - D_X^N Y$ be the second fundamental of N in M. Then

$$S(X,JY) = D_X^M(JY) - D_X^n(JY) = J_M D_X^M Y - J_N D_X^N Y$$

$$= J_M(D_X^M Y - D_X^N Y) = JS(X,Y) .$$

By symmetry,

$$S(JX,Y) = S(Y,JX) = JS(Y,X) = JS(X,Y) .$$

For any submanifold, the sectional curvatures satisfy (X,Y orthonormal, P = span of X,Y):

$$K_N(P) = R^M(X,Y,X,Y) + g_M(S(X,X),S(Y,Y)) - g_M(S(X,Y),S(X,Y))$$

$$= K_M(P) + g_M(S(X,X),S(Y,Y)) - g_M(S(X,Y),S(X,Y)) .$$

In our case, if $Y = JX$ then

$$K_N(P) = K_M(P) + g_M(S(X,X),S(JX,JX)) - g_M(S(X,JX),S(X,JX))$$

$$= K_M(P) - 2g(S(X,X),S(X,X))$$

because

$$S(JX,JX) = - S(X,X)$$

and

$$g_M(S(X,JX),S(X,JX)) = g_M(JS(X,X),JS(X,X)) = g_M(S(X,X),S(X,X)) .$$

In particular, the holomorphic sectional curvature of N is always $\leq$ the corresponding sectional curvature of M. Let us compute the trace of S: we can of course compute relative to any basis so we use a basis of the form $X_1, JX_1, ... X_m, JX_m$, $m = \dim_{\mathbb{C}} N$.

Then

$$\text{Trace } S = S(X_1, X_1) + S(JX_1, JX_1) + \cdots + S(X_n, X_n) + S(JX_n, JX_n) = 0$$

because $S(JX_1, JX_1) = -S(X_1, X_1)$. So N is a minimal submanifold of M. (In case N is compact and so is M, it can be shown that N is absolutely area minimizing in the homology class it represents in M, this homology class being necessarily nontrivial if N is not zero dimensional.)

11. Holomorphic Vector Bundles and Hermitian Metrics and Connections

A holomorphic vector bundle is defined just as is a topological vector bundle with fibres **C**-vector spaces with an additional restriction, that the transition functions be holomorphic functions of the point in the manifold. More explicitly, with $B \xrightarrow{\pi} M$, M a complex manifold bundle, we suppose given a trivilizing cover U_λ and maps:

$$\phi_\lambda : \pi^{-1}(U_\lambda) \to U_\lambda \times \mathbf{C}^k$$

with $\pi_1 \circ \phi_\lambda = \pi/_{\pi^{-1}(U_\lambda)}$ when $\pi_1 =$ first factor projection and the linear maps, defined for $p \in U_{\lambda_1} \cap U_{\lambda_2}$ of $\mathbf{C}^k \to \mathbf{C}^k$ by $\pi_2 \circ \phi_{\lambda_2} \circ (\phi_{\lambda_1}^{-1} | p \times \mathbf{C}^k)$ to depend holomorphically on p. (Notes: $\pi_2 =$ projection on the second factor; since linear maps $\mathbf{C}^k \to \mathbf{C}^k$ are uniquely associated to $k \times k$ **C**-valued matrices, it makes sense to speak of such maps being holomorphic: it just means that each matrix element is a holomorphic function.)

Notation: $\pi_2 \circ \phi_{\lambda_2} \circ (\phi_{\lambda_1}^{-1} | p \times \mathbf{C}^k) = f_{\lambda_1 \lambda_2} \in \mathbf{C}^{k^2}$.

A *Hermitian metric* on a holomorphic vector bundle $B \xrightarrow{\pi} M$ is a C^∞ family of Hermitian metric (in the standard linear algebra sense) on each fibre $\pi^{-1}(p)$, $p \in M$.

A **R**-vector bundle with Riemannian metric in general admits a wide variety of metric-preserving connections, i.e., connections for which parallel translation preserves inner product. (Recall: The unique Riemannian connection on TM, M a Riemannian manifold is made unique only by imposing the additional condition of torsion 0.) Similarly, a complex Hermitian vector bundle admits many Hermitian-metric preserving connections. In the case of a holomorphic Hermitian vector bundle, however, there is a natural way to select a unique metric-preserving connection from among the many possible metric-preserving connections.

DEFINITION. A connection on a holomorphic vector bundle $B \xrightarrow{\pi} M$ is type (1,0) if its connection forms relative to a local holomorphic frame in B are type (1,0).

It is easy to check that the definition does not depend on which holomorphic local frame is used.

The basic uniqueness and existence result is the following:

THEOREM. *If $B \xrightarrow{\pi} M$ is a holomorphic vector bundle and if h is a Hermitian (fibre) metric on B, then $\exists$ a unique type (1,0) connection on B that is metric preserving.*

Proof. Choose a local holomorphic frame, i.e., a trivialization
$\phi : \pi^{-1}(U) \to U \times \mathbf{C}^k$ and set $\sigma_j = \phi^{-1}((0 \cdots 1 \cdots 0)$ so that $\sigma_1,...,\sigma_k$ are holomorphic and span $\pi^{-1}(p)$ at each $p \in U$. Set

$$h_{\alpha\beta} = h(\sigma_\alpha, \sigma_\beta) \quad 1 \le \alpha \le k, 1 \le \beta \le k \; .$$

Define the connection forms of a covariant differentiation (connection) D to be those determined by

$$D\sigma_\alpha = \sum_{\beta=1}^{k} \omega_\alpha^\beta \, \sigma_\beta, \quad \alpha = 1,...,k \; .$$

(i.e.

$$D_X\sigma_\alpha = \sum_{\beta=1}^{k} \omega_\alpha^\beta(X) \, \sigma_\beta) \; .$$

Then D is metric-preserving if and only if

$$dh_{\alpha\beta} = h(D\sigma_\alpha, \sigma_\beta) + h(\sigma_\alpha, D\sigma_\beta)$$

or

$$dh_{\alpha\beta} = h\left(\sum_{\gamma=1}^{k} \omega_\alpha^\gamma \sigma_\gamma, \sigma_\beta\right) + h\left(\sigma_\alpha, \sum_{\delta=1}^{k} \omega_\beta^\delta \sigma_\delta\right)$$

$$= \sum_{\gamma=1}^{k} \omega_\alpha^\gamma h(\sigma_\gamma, \sigma_\beta) + \sum_{\delta=1}^{k} \overline{\omega}_\beta^\delta h(\sigma_\alpha, \sigma_\delta).$$

(Note that the $\overline{\omega}$ is conjugate in the second sum because h is conjugate linear in the second variable.)

If ω's are type (1,0) so that $\overline{\omega}$'s are type (0,1) then we must have

(†) $$\partial h_{\alpha\beta} = \sum_{\gamma=1}^{k} \omega_\alpha^\gamma h_{\gamma\beta}$$

and

$$\overline{\partial} h_{\alpha\beta} = \sum_{\delta=1}^{k} \overline{\omega}_\beta^\delta h_{\alpha\delta}.$$

If the first set of these equations hold, then so do the second because

$$\overline{\partial} h_{\alpha\beta} = \overline{\partial h_{\overline{\alpha\beta}}} = \overline{\partial h_{\beta\alpha}} = \overline{\left(\sum_{\delta=1}^{k} \omega_\beta^\delta h_{\delta\alpha}\right)}$$

$$= \sum_{\delta=1}^{k} \overline{\omega}_\beta^\delta \overline{h}_{\delta\alpha} = \sum_{\delta=1}^{k} \overline{\omega}_\beta^\delta h_{\alpha\delta}, \text{ as required }.$$

On the other hand, the fact that the matrix $(h_{\gamma\beta})$ is invertible (being positive definite) means that we can choose the ω_α^γ in one and only one way so as to make the first equations (†) work. So we have local existence, and uniqueness for the desired connection. Global existence (and uniqueness) follows as usual. ∎

In a holomorphic frame, we have from the previous

$$\omega_\alpha^\delta = \sum_\beta h^{\beta\delta} \partial h_{\alpha\beta}$$

where $h^{\beta\delta}$ is the inverse of h (so $\sum_\lambda h_{\rho\lambda} h^{\lambda\mu} = \delta_\rho^\mu$).

An interesting special case is that of holomorphic line bundles. (Note: Line bundles in the **C** sense are more interesting than those in the **R**-sense. The latter have discrete (± 1) structure group, i.e., are reducible to that group. But in general **C** bundles are reducible only as far as $S^1 \subset \mathbf{C}^*$, not to locally constant transition functions.)

In the line bundle case, h is a 1×1 matrix. Also

$$\omega_1^1 = 1/h_{11}\, \partial h_{11} = \partial(\log h_{11})$$

(for log as in the real sense: $h_{11} > 0$). In a different frame, h_{11} changes to $f\bar{f}\, h_{11}$, f a (nonvanishing) holomorphic function. It follows that

$$\bar{\partial}\partial(\log h_{11} f\bar{f}) = \bar{\partial}\partial(\log h_{11} + \log f + \log \bar{f}) = \bar{\partial}\partial \log h_{11},$$

where $\log f$ and $\log \bar{f}$ can be any fixed local branch of (holomorphic) "log". So

$$\bar{\partial}\partial(\log h_{11})$$

is in fact a globally defined type (1,1) form.

Associated algebraically to the well-defined form $\bar{\partial}\partial \log h_{11}$ is the Hermitian form

(††) $$-\sum_{i,j=1}^n \left[\frac{\partial^2}{\partial z_i \partial \bar{z}_j} \log h_{11}\right] dz_i \otimes d\bar{z}_j .$$

We shall call this form the Hermitian curvature form. The association is essentially that of metric to Kähler form. But to avoid algebraic detail, it is more convenient simply to compute directly that the Hermitian form is independent of local trivialization. This is easy following the line of reasoning used to show that $\bar{\partial}\partial \log h_{11}$ is well defined, and it is left to the reader. (This approach also avoids the possibility of sign errors, which have plagued the transition from type (1,1) to Hermitian forms in the literature.)

Special interest is attached to Hermitian holomorphic line bundles for which the Hermitian curvature form indicated is definite, either positive everywhere or negative everywhere. We make a formal definition:

DEFINITION: A holomorphic line bundle B is *positive* (respectively, *nonnegative*) if for some Hermitian metric on B the Hermitian form (††) is positive definite (respectively, nonnegative definite). The bundle B is *negative* (respectively, *nonpositive*) if for some Hermitian metric on B the Hermitian curvature form is everywhere negative definite (respectively, nonpositive definite).

The logic of the negative sign in the definition of the Hermitian curvature form is as follows. Conventionally, it has been the practice to regard bundles with global holomorphic sections positively, having sections being a good property. Now if a Hermitian line bundle over a compact complex manifold has a nontrivial holomorphic section s then for a local frame field σ_1 we can write $s = f\sigma_1, f$ holomorphic, and then, where $s \neq 0$:

$$\sum_{i,j=1}^{n} \left[\frac{\partial^2}{\partial z_i \partial \bar{z}_j} \log h_{11} \right] dz_i \otimes d\bar{z}_j = \sum_{i,j=1}^{n} \left[\frac{\partial^2}{\partial z_i \partial \bar{z}_j} \log(h_{11} f \bar{f}) \right] dz_i \otimes d\bar{z}_j$$

$$= \sum_{i,j=1}^{n} \left[\frac{\partial^2}{\partial z_i \partial \bar{z}_j} \log \|s\|^2 \right] dz_i \otimes d\bar{z}_j .$$

(Here $\|s\|^2$ is the Hermitian norm squared of the section s.) The last-written form must be nonpositive definite at the point(s) where the global function $\|s\|^2$ attains its maximum. (Exercise in calculus: Prove this.) In particular, there must be points of M where the Hermitian form

$$\sum_{i,j=1}^{n} \left[\frac{\partial^2}{\partial z_i \partial \bar{z}_j} \log h_{11} \right] dz_i \otimes d\bar{z}_j$$

must be nonpositive definite or equivalently points at which

$$-\sum_{i,j=1}^{n} \left[\frac{\partial^2}{\partial z_i \partial \bar{z}_j} \log h_{11} \right] dz_i \otimes d\bar{z}_j$$

is nonnegative definite.

Note that no such reasoning occurs at the minimum of $\|s\|^2$ in general because the minimum may well be zero, in which case $\log \|s\|^2$ is not defined at the minimum. What we do obtain by following the above pattern is that if s is a holomorphic section which nowhere vanishes (so that minimum $\|s\|^2 > 0$ and $\log \|s\|^2$ is defined and C^∞ globally) then the Hermitian curvature form must be nonpositive definite somewhere. Of course, such a section exists if and only if B is the trivial line bundle. Thus we see that a line bundle on a compact complex manifold that is positive or negative cannot be trivial.

Every complex manifold M has a naturally arising holomorphic line bundle on it. This is the bundle of forms of type $(n,0)$; this bundle is called the *canonical bundle* of M and denoted by K or, when the manifold needs specification, K_M.

If M has a Hermitian metric g then K_M can be given an associated Hermitian metric as follows: Let $(z_1,\ldots,z_n)$ be a local coordinate system on M and set $g_{i\bar{j}} = g(\partial/\partial z_i, \partial/\partial \bar{z}_j)$ as usual. Then put

$$\|dz_1 \wedge \cdots \wedge dz_n\|^2 = 1/\det(g_{i\bar{j}}),$$

where $\det(g_{i\bar{j}}) =$ the determinant of the matrix $(g_{i\bar{j}}), 1 \le i,j \le n$. A slightly tedious but routine calculation shows that this definition yields a well defined C^∞ Hermitian metric on K_M. We shall call this the canonical metric on K_M (associated to the metric g on M).

The Hermitian curvature form of the canonical metric on K_M is closely related to the Riemannian curvature tensor for a Kähler manifold M. To make this relationship explicit, let R be the Riemannian curvature (4-) tensor of the Kähler manifold M. Define, as usual, the Ricci tensor Ric of M by (at each point of M)

$$\text{Ric}(X,Y) = \sum_{i=1}^{2n} R(X,e_i,Y,e_i),$$

where $\{e_i | i=1,\ldots 2n\}$ is an orthonormal (real) basis for the real tangent space M_p of M at p and $X,Y \in M_p$. It is easy to check that $\text{Ric}(X,Y)$ is independent of the choice of orthonormal basis. It is also easy to see, using the J-symmetries of the curvature tensor, that Ric is an Hermitian form, i.e.,

$$\text{Ric}(JX,JY) = \text{Ric}(X,Y) .$$

The Hermitian real form Ric extends by complex linearity to $M_p \otimes C$. On the holomorphic tangent space of M at p we define a C-Hermitian form by a now familiar process:

$$(Z,W) \to \text{Ric}(Z,\overline{W}) .$$

The promised relationship between the Hermitian curvature of K_M and the curvature of M is given now by the formula (with $G = \det(g_{ij})$, $1 \le i,j \le n$):

$$\left[\sum_{i,j=1}^{n} \frac{\partial^2}{\partial z_i \partial \overline{z}_j} (\log G) \, dz_i \otimes d\overline{z}_j \right] (Z,\overline{W}) = -\text{Ric}(Z,\overline{W})$$

for all Z,W in the holomorphic tangent space of M. Here the left hand side is exactly the Hermitian curvature form of K_M evaluated on $Z,\overline{W}$ because the metric of K_M is (locally) $1/G$ and $\log(1/G) = -\log G$. Thus, for example, if M has positive sectional curvature and hence positive definite (real) Ricci form, it follows that K_M is a negative bundle. Specific examples of this: P_nC, all n. The formula just displayed can be proven by a tedious but straightforward calculation, using the standard rules for differentiation of determinants.

Note: In case M has real dimension 2, then the formula reduces to the familiar expression for the Gauss curvature for a metric of the form $G(dx^2 + dy^2)$, namely $k = -1/2 \triangle (\log G)$.

We can see that the signs check out correctly in our relationship between Hermitian curvature of the canonical bundle and the Ricci curvature of the manifold by considering complex projective space. It has positive sectional curvature and hence positive Ricci curvature, in its usual Fubini-Study Kähler metric. Thus, according to our calculation, its canonical bundle K is negative (in its canonical metric). Negative bundles cannot have nontrivial holomorphic sections, and sure enough there are no nontrivial holomorphic $(n,0)$ forms on complex projective n-space $P_n\mathbb{C}$.

Looking at this same example the other way around, let us consider K_M^*, the dual bundle of K_M. We can put on K_M^* the dual of the canonical metric, namely, if σ_1 is a local nonvanishing section of K_M^* and σ_2 is a local nonvanishing section of K_M we set

$$\|\sigma_1\|^2 = |\sigma_1(\sigma_2)|/\|\sigma_2\|^2 .$$

It is easy to see that the Hermitian curvature of K_M^* in this metric is just $-1\times$ the Hermitian curvature of K_M. (This hold thing works for any line bundle and its dual: exercise.) Thus K_M^* is a positive bundle. This is as it should be since K_M^* has nontrivial global holomorphic sections (obtained by wedging together holomorphic vector fields). Exercise: Figure out one such section on $P_1\mathbb{C} \cong$ the Riemann sphere.

We have already noted that line bundles with a nontrivial holomorphic section must have positive Hermitian curvature form somewhere (in any Hermitian metric). There is a profound and important converse of this, first established by K. Kodaira.

By definition, the tensor product $B_1 \otimes B_2$ of two line bundles is the line bundle whose transition functions are the products of the transistion functions of B_1 and B_2. (Exercise: $\otimes$ on bundles corresponds to addition of Hermitian curvature forms. Formulate this precisely and prove it.) Along these same lines, it can be shown that sufficiently high $\otimes$-powers actually give an embedding into projective space in the following sense: Let $B \to M$ be a line bundle over a compact complex manifold M. Let $H^\circ(B) =$ the C-vector space of holomorphic sections of B. This is actually a finite-dimensional vector space. (It is easy to see, for instance, that its unit ball is compact in the norm $\|s\|_M = (\int_M \|s\|^2)^{1/2}$ where $\|s\|$ is the pointwise norm in an arbitrary Hermitian metric on B and the integral over M is relative to the volume form of an arbitrary Hermitian metric on M).

Choose a basis $s_1,...,s_k$ for $H^°(B)$ and suppose (additionally! - this doesn't have to happen) that there is no point $x \in M$ with all $s_j(x)=0$, $j=1,...,k$. Then we can define a map $\mathcal{E}_B : M \to P_{k-1}\mathbf{C}$ as follows. For $x \in M$, choose a $j \in \{1,...,n\}$ with $s_j(x) \neq 0$. Then set the image $\mathcal{E}_B(x)$ of $x=$ the point with homogeneous coordinates $s_1(x)/s_j(x),...,1,...,s_k(x)/s_j(x)$, where the 1 is in the jth slot. That is, the image of x is the image of $(s_1(x)/s_j(x),...,1,...,s_k(x)/s_j(x))$ $\in \mathbf{C}^k - \{(0,...0)\}$, under the usual quotient map $\mathbf{C}^k - \{(0,...,0)\} \to P_{k-1}\mathbf{C}$. The map $\mathcal{E}_B$ is holomorphic. A different choice of the basis s_j changes $\mathcal{E}_B$ only by a linear isomorphism $P_n\mathbf{C}$.

Then the following important theorem holds, the famous Kodaira Embedding Theorem:

If M is a compact complex manifold and if $B \to M$ is a positive line bundle, then there is a positive integer ℓ such that $\mathcal{E}_{B^\ell}$ is defined and is a holomorphic embedding of M into a complex projective space.

As noted, $P_n\mathbf{C}$ has a positive holomorphic line bundle on it, namely K^*. If M is a compact complex submanifold of $P_n\mathbf{C}$ then it is easy to see that $K^*|M$ is positive. (In fact, the restriction of a positive bundle is positive: prove this as an exercise.) Thus we see from the Kodaira Embedding Theorem that a compact complex manifold is biholomorphic to a submanifold of some complex manifold if and only if M has on it a positive holomorphic line bundle.

The proof of the Kodaira Theorem is highly nontrivial and will not be given here.

In case that the canonical bundle is positive or negative or a trivial bundle, it is natural to ask whether it is possible to make the Hermitian curvature form of the canonical bundle equal to a scalar multiple of the metric itself. In particular, one could ask whether there is a Kähler metric g on M such that, for some $c \in \mathbf{R}$,

$$\text{Ric}(Z,\overline{W}) = cg(Z,\overline{W})$$

for all (complex) vectors $W \in M_p \otimes \mathbf{C}$, all $p \in M$. In this set up, $c < 0$ corresponds to K_M positive, $c = 0$ to K_M trivial, and $c > 0$ to K_M negative. This is known not to be possible in general if K_M is negative. But if K_M is trivial or positive then in fact such Kähler metrics always exist. This important and deep result was shown by S. T Yau after an earlier conjecture by E. Calabi. (The case $c < 0$ was established independently of Yau's work by T. Aubin.) These results are discussed in detail in reference [1].

Another particulary interesting holomorphic vector bundle is the *holomorphic tangent bundle* of a complex manifold M, i.e., the bundle with fibre at $p =$ by definition to the holomorphic tangent space at $p =$ by definition the C-linear span of $\partial/\partial z_1,\ldots, \partial/\partial z_n$, $(z_1,\ldots,z_n)$ a holomorphic coordinate system near p. Suppose M is given a Hermitian metric g, i.e., a J-invariant Riemannian metric. Then $T^h M$, The holomorphic tangent bundle of M, becomes a Hermitian holomorphic vector bundle, with the Hermitian metric determined by g (as in an earlier section). The Kähler case of this situation is especially nice, as we shall see after we prove the following lemma.

LEMMA. *Suppose $B \xrightarrow{\pi} M$ is a Hermitian holomorphic vector bundle and $\Sigma_1,\ldots,\Sigma_k$ is a basis for the fibre B_p at a point $p \in M$. Then $\exists$ a local holomorphic frame $\sigma_1,\ldots,\sigma_k \ni$*

(1) $$\sigma_j|_p = \Sigma_j, \quad \text{all } j = 1,\ldots,k$$

and

(2) $$D\sigma_j|_p = 0, \quad \text{all } j = 1,\ldots,k .$$

Proof. Choose a local holomorphic frame $\gamma_1,\ldots,\gamma_k$ with $\gamma_j|_p = \Sigma_j$, all $j = 1,\ldots,k$. Now compute

$$D(\sum_\beta f^{\alpha\beta} \gamma_\beta) = \sum_\beta df^{\alpha\beta} \gamma_\beta + \sum_{\beta,\delta} f^{\alpha\beta} \omega_\beta^\delta \gamma_\delta .$$

This $= 0$ at p if and only if

$$df^{\alpha\beta} = - \text{ the } \gamma_\beta \text{ component of } \sum_{\beta,\delta} f^{\alpha\beta} \omega_\beta^\delta \gamma_\delta$$

$$= - \sum_\rho f^{\alpha\rho} \omega_\rho^\beta .$$

Since this last form is type (1,0) we can choose a set of holomorphic $f^{\alpha\beta}$ with $f^{\alpha\beta}|_p =$ and $df^{\alpha\beta}$ to satisfy the equations just given. Then $D(\sum_\beta f^{\alpha\beta} \gamma_\beta)|_p = 0$ all $\alpha = 1,\ldots,k$. Since f is then invertible near p,

$$\sigma_\alpha = \sum_\beta f^{\alpha\beta} \gamma_\beta$$

fits the requirements. ∎

If M is a Hermitian a manifold (i.e. M has a J-invariant Riemannian metric attached) then the Riemannian connection on M induces by complex linear extension a connection on $TM \otimes \mathbf{C}$, the complexified tangent bundle of M. If M is Kähler, then T^hM is a parallel subbundle of $TM \otimes \mathbf{C}$. (Here a subbundle is *parallel* by definition if parallel translation of a vector in the subbundle remains in the subbundle.) This follows because J is parallel so its eigenspaces are preserved by parallel translation.

Moreover, again if M is Kähler, the Riemannian connection on T^hM determined by considering T^hM as a Hermitian vector bundle and finding its unique type (1,0) connection. The proof of this fact is easy: Choose a holomorphic normal coordinate system at a point p, say $(z_1,\ldots,z_n)$. Then for the Riemannian connection D

$$D \frac{\partial}{\partial z_j}\bigg|_p = 0 \quad j = 1,\ldots,n$$

by definition. Let $\nabla =$ the Hermitian connection for T^hM. Then also

$$\nabla \frac{\partial}{\partial z_j}\bigg|_p = 0 \quad j = 1,\ldots,n .$$

because

$$dg\left(\frac{\partial}{\partial z_i}, \frac{\bar\partial}{\partial z_j}\right)\bigg|_p = 0 .$$

So $D = \nabla$.

The converse of this is also true. For this, let M be a Hermitian manifold – with Riemannian metric (J-invariant) $= g$ and Riemannian metric (J-invariant) $= g$ and Riemannian connection D. Then D induces a connection $\hat{D}$ on T^hM by setting

$$\hat{D}_z W = \text{the projection of } D_z W \text{ on } T^hM ,$$

where projection is relative to the *canonical* splitting $TM_p \otimes \mathbf{C} = T^hM_p \oplus \overline{T^hM_p}$. (This also happens to be orthogonal projection relative to g extended to $TM \otimes \mathbf{C}$ as a Hermitian metric.) In this setting, we can see that if $\hat{D} = \nabla$ ($\nabla \stackrel{\text{def}}{=}$ the Hermitian connection) then M is Kähler. [Note: On the previous pages, we showed that if M is Kähler, then $\hat{D} = D = \nabla$.]

To prove this, choose a local holomorphic frame σ_j in T^hM which is orthonormal at p and has $\nabla = 0$ at p (by the Lemma). Then, since $\nabla = \hat{D}$ by hypothesis, we have

$$\hat{D}\sigma_j\big|_p = 0 .$$

But $\hat{D}$ is a length compatible connection [i.e. $g(\hat{D}_X Z, \overline{W}) + g(Z, \hat{D}_X W) = Xg(Z,\overline{W})$] because D is and the projection involved in $\hat{D}$ is orthogonal. So

$$dg(\sigma_j, \overline{\sigma}_k)\big|_p = 0 \quad j,k = 1,\ldots,n .$$

Moreover, $\hat{D}$ has torsion 0 because D does and Lie brackets of vector fields preserves types, i.e., $[Z,W] \in T^hM$ if $Z,W \in T^hM$. Since $\hat{D}\sigma_j\big|_p = 0$, it follows that $[\sigma_j, \sigma_k] = 0$ at p, all j, k. Now given a holomorphic local frame σ_j with $[\sigma_j, \sigma_k]\big|_p = 0$, there is a holomorphic coordinate system $(z_1,\ldots,z_n)$ such that $\sigma_j\big|_p = \partial/\partial z_j\big|_p$ and moreover $\sigma_j - \partial/\partial z_j$ vanishes to order two at p. Then (z_j) will be (in our case) a holomorphic normal coordinate system at p. So M is Kähler. ∎

Appendix: Integrable and Nonintegrable Almost Complex Structures

For convenience, we based our decomposition of differential forms into types (p,q) on the use of complex (holomorphic) coordinates $(z_1,\ldots,z_n)$: i.e. $\{(p,q) \text{ forms}\}$ = span of $dz_{i_1} \wedge \cdots \wedge dz_{i_p} \wedge d\bar{z}_{j_1} \wedge \cdots \wedge d\bar{z}_{j_q}$. However, it is important to realize that this was only for convenience. The whole question of types of forms is after all a strictly pointwise item, and one should expect to be able to consider it algebraically. Specifically, we can consider it as follows:

Let V be a real $2n$-dimensional vector space with an endomorphism $J: V \to V \ni J^2 = -1$. Then $V^* \otimes \mathbb{C}$ = direct sum of two subspaces determined as follows: Let $X_1, JX_1, \ldots X_n, JX_n$ be a basis for V and $\omega_1, \omega_2 \cdots \omega_{2n-1}, \omega_{2n}$ be the dual basis. [So if $X_1 = \partial/\partial x_1, \ldots, \omega_1 = dx_1, \omega_2 = dy_1$ etc.]. Then

$$V^* \otimes \mathbb{C} = \text{span}\{\omega_{2k-1} + \sqrt{-1}\omega_{2k}, \quad k = 1,\ldots,n\}$$

$$\oplus \text{span}\{\omega_{2k-1} - \sqrt{-1}\omega_{2k}, k = 1,\ldots,n\}.$$

This decomposition is easily checked to be invariant under changes of the $X_1, JX_1, \ldots, X_n, JX_n$ basis (to another of the same form). This invariance can be in fact checked without computation as follows: The span of $X_1 + \sqrt{-1} JX_1, \ldots, X_n + \sqrt{-1} X_n$ which corresponds to $\partial/\partial \bar{z}$'s is the $-\sqrt{-1}$ eigenspace of J extended by complex linearity to $V^* \otimes \mathbb{C}$: e.g.,

$$J(X_1 + \sqrt{-1} JX_1) = \sqrt{-1} J^2 X_1 + JX_1 = -\sqrt{-1} X_1 + JX_1$$

$$= -\sqrt{-1}(X_1 + \sqrt{-1} JX_1).$$

Similarly

$$J(X_1 - \sqrt{-1} JX_1) = -\sqrt{-1} J^2 X_1 + JX_1 = \sqrt{-1} X_1 + JX_1$$

$$= \sqrt{-1}(X_1 - \sqrt{-1} JX).$$

So

$$V \otimes \mathbb{C} = (-\sqrt{-1}) \text{ eigenspace of } J \dotplus \sqrt{-1} \text{ eigenspace of } J.$$

The decomposition of $V^* \otimes \mathbf{C}$ is then determined by

$$V^{1,0} \stackrel{\text{det}}{=} \text{span}\{\omega_{2k-1} + \sqrt{-1}\,\omega_{2k}, \ k = 1,\ldots,n\}$$

$$= \{\omega \in V^* \otimes \mathbf{C} \mid \omega(X) = 0, \text{ all } X \in$$
$$- \sqrt{-1} \text{ eigenspace of } J \text{ on } V \otimes \mathbf{C}\}$$

[e.g. $dz_1((\partial/\partial x_1) + \sqrt{-1}\,(\partial/\partial y_1)) = 0$].

$$V^{0,1} \stackrel{\text{def}}{=} \text{span}\{\omega_{2k-1} + \sqrt{-1}\,\omega_{2k}, k = 1,\ldots,n\}$$

$$= \{\omega \in V^* \otimes \mathbf{C} \mid \omega(x) = 0, \text{ all } X \in$$
$$+ \sqrt{-1} \text{ eigenspace of } J \text{ on } V \otimes \mathbf{C}\}.$$

[e.g. $d\bar{z}_1((\partial/\partial x_1) - \sqrt{-1}\,(\partial/\partial y_1)) = 0$].

Thus $V^* \otimes = V^{1,0} \oplus V^{0,1}$ and this is basis-choice independent.

This decomposition of $V^* \otimes \mathbf{C}$ gives rise to a decomposition of the whole complex exterior algebra (over $\mathbf{C}$) $\Lambda_{\mathbf{C}}(V^* \otimes \mathbf{C})$ and hence to a type (p,q) decomposition as before.

Now suppose M is a C^∞ manifold on which there is a C^∞ family of endomorphisms $J_p : M_p \to M_p$, $p \in M$, $\ni. \ J_p^2 = -1$ for all $p \in M$. Such a family is called an *almost complex structure* on M. (Note: Not every manifold admits an almost complex structure.) We specifically now do *not* assume that the almost complex structure J comes from a complex structure on M, i.e., a covering of M by coordinate charts with holomorphic overlaps. The question then naturally arises, when does such a general almost complex structure in fact come from a complex structure? Necessary conditions are not hard to find:

Let $\mathcal{F}^{p,q}$ = the set of C^∞ $\mathbf{C}$-valued differential forms ω on $M \ni \omega|_p$ is type (p,q) at p for all $p \in M$. Then $\mathcal{F}^{0,0}$ = $\mathbf{C}$-valued functions on M, etc. Note that

$\mathcal{F}^{0,0}$, $\mathcal{F}^{1,0}$ and $\mathcal{F}^{0,1}$ generate (under $\wedge$) locally $\mathcal{F} = \cup_{p,q} \mathcal{F}^{p,q}$. Also

$$d\mathcal{F}^{0,0} \subset \mathcal{F}^{1,0} \oplus \mathcal{F}^{0,1} = \text{all 1-forms}$$

$$d\mathcal{F}^{1,0} \subset \mathcal{F}^{1,1} \oplus \mathcal{F}^{2,0} \oplus \mathcal{F}^{0,2} = \text{all 2-forms}$$

$$d\mathcal{F}^{0,1} \subset \mathcal{F}^{1,1} \oplus \mathcal{F}^{2,0} \oplus \mathcal{F}^{0,2}$$

It follows (from the local generation of $\mathcal{F}$) that

$$d\mathcal{F}^{p,q} \subset \mathcal{F}^{p+1,q} \oplus \mathcal{F}^{p,q+1} \oplus \mathcal{F}^{p+2,q-1} \oplus \mathcal{F}^{p-1,q+2}.$$

For a complex structure

$$d\mathcal{F}^{p,q} \subset \mathcal{F}^{p+1,q} \oplus \mathcal{F}^{p,q+1}.$$

So a *necessary condition* for an almost complex structure J to come from a complex structure is that

(*) $$d\mathcal{F}^{p,q} \subset \mathcal{F}^{p+1,q} \oplus \mathcal{F}^{p,q+1}$$

when $\mathcal{F}$ is decomposed according to J types.

(Note: If $\dim_R M = 2$, then $\mathcal{F}^{p,q} = 0$ if $p \geq 2$ or $q \geq 2$. Hence the necessary condition (*) is automatically satisfied.)

It is a fact — not easily proved, however — that the necessary condition (*) is in fact sufficient. An almost complex structure satisfying (*) is said to be *integrable*. Then we have Theorem (Newlander-Nirenberg): If J is a C^∞ integrable almost complex structure on a C^∞ manifold M, then there is a complex structure on M such that the J-mapping associated to this complex structure = the given almost complex structure J.

It is of interest to look at what this theorem says in the $\dim_R M = 2$ case: J determines an orientation on M. Moreover, we can construct a J-invariant metric (Riemannian) on M by setting

$$g(X,Y) = G(X,Y) + G(JX,JY)$$

where G is an arbitrary Riemannian metric on M. Now if $z = x + iy$ is a J-holomorphic coordinate system on M i.e. if $J(\partial/\partial x) = \partial/\partial y$ then the metric g is given by

$$\lambda^2(dx^2 + dy^2)$$

where λ is a C^∞ function. The converse also holds: If (x,y) is an oriented real coordinate system and if $g(\partial/\partial x, \partial/\partial x) = g(\partial/\partial y, \partial/\partial y)$ and $g(\partial/\partial x, \partial/\partial y) = 0$, then $x + iy$ is a holomorphic coordinate on M, i.e., $J(\partial/\partial x) = \partial/\partial y$. Thus the problem of finding a complex structure on M compatible with the given almost complex structure J is equivalent to finding real local coordinates (x,y) such that $g(\partial/\partial x, \partial/\partial x) = g(\partial/\partial y, \partial/\partial y)$ and $g(\partial/\partial x, \partial/\partial y) = 0$, i.e., so called "isothermal coordinates". This problem can always be solved, for any C^∞ Riemannian metric g (partial differential equation methods). In our setting, this solvability corresponds to the fact that an almost complex structure on a manifold M with $\dim_R M = 2$ is always integrable.

There are other conditions that are equivalent to integrability of an almost complex structure. One of the more useful of these is as follows:

Define a tensor field N by

$$N(X,Y) = 2\{[JX,JY] - [X,Y] - J[X,JY] - J[JX,Y]\},$$

X, Y real vector fields. A calculation shows that N is function-linear in X and Y and hence that N is a (pointwise) tensor. If there is a complex J-holomorphic coordinate system defined in a neighborhood of $p \in M$ (i.e., a coordinate system $x_1 + \sqrt{-1}\, y_1, \ldots, x_n + \sqrt{-1}\, y_n \ni J(\partial/\partial x_j) = \partial/\partial y_j$), then

$$N_p = 0$$

because by inspection N has 0 components in the given coordinates. Thus the J of a complex structure has $N = 0$.

Another condition that we could impose on J has to do with "vector fields of type (1,0) or type (0,1)". We define a $\mathbf{C}$-vector field. (i.e., a vector field X with $X_p \in M_p \otimes \mathbf{C}$ for each $p \in M$) to be of type

(1,0) if $JX = \sqrt{-1}\, X$

(0,1) if $JX = -\sqrt{-1}\, X$.

The condition for the integrability of J becomes that type (1,0) vector fields are a Lie subalgebra, i.e. $[Z,W]$ is type (1,0) if z and W are of type (1,0). Since $[Z,W] = [\bar{Z},\bar{W}]$, this condition is equivalent to type (0,1) vector fields being a Lie subalgebra. It can be shown that either of these (trivially) equivalent conditions, as well as $N = 0$, is equivalent to integrability in our previous sense and hence equivalent to the existence of an associated complex structure.

References

The journal literature of complex differential geometry is far too large to admit a quick summary or even a brief representative sampling. The following books will enable the reader to continue further with the topics introduced here, and will provide an introduction to the remainder of the literature.

Items [4] and [6] cover in detail the basic results about holomorphic functions of several complex variables, and a great deal beyond the basic results.

Chapter IX of item [5] is a treatment of basic complex differential geometry much along the lines of the discussion here; an appendix to [5] contains the proof of the integribility theorem for complex structures in the real analytic case; [5] also has a considerable bibliography up to its publication date. Item [2] gives a treatment of complex manifolds with a somewhat different emphasis involving a more cohomological viewpoint including characteristic classes, etc.

Item [9], though not specifically directed toward complex differential geometry, will provide some indications of recent developments in the field.

Item [1] provides a detailed treatment of the most important recent development in the field, S. T. Yau's proof of the Calabi conjecture.

The algebraic geometric viewpoint on complex manifolds is discussed extensively in [3]; this includes the proof of the fundamental Kodaira theorems mentioned here in §11. The Kodaira theorems are also treated in [7] and [8].

1. B. Bergery, J. P. Bourguignon, E. Calabi, S. T. Yau, et. al., *Premiere Classe de Chern et Courbure de Ricci: Preuve de la Conjecture de Calabi*. Seminaire Palaiseau 1978. Societe Mathematique de France, Asterisque 58. Paris. 1978.

2. S. S. Chern, *Complex Manifolds without Potential Theory*, second edition, Springer-Verlag, New York, 1979.

3. P. Griffiths and J. Harris, *Principles of Algebraic Geometry*, John Wiley, New York, 1978.

4. L. Hörmander, *An Introduction to Complex Analysis in Several Variables*, second edition, North Holland-American Elsevier, New York, 1973.

5. S. Kobayashi and K. Nomizu, *Foundations of Differential Geometry*, John Wiley, New York, 1969.

6. S. Krantz, *Function Theory of Several Complex Variables*, John Wiley, New York, 1982.

7. J. Morrow and K. Kodaira, *Complex Manifolds*, Holt, Rinehart, Winston. New York, 1971.

8. R. Wells, *Differential Analysis on Complex Manifolds*, second edition, Springer-Verlag. New York, 1980.

9. S. T. Yau, *Seminar on Differential Geometry*, Princeton University Press, Annals of Mathematics Studies, No. 102, Princeton. 1982.

MIX
Papier aus verantwortungsvollen Quellen
Paper from responsible sources
FSC® C105338

If you have any concerns about our products,
you can contact us on
ProductSafety@springernature.com

In case Publisher is established outside the EU,
the EU authorized representative is:
Springer Nature Customer Service Center GmbH
Europaplatz 3, 69115 Heidelberg, Germany

Printed by Libri Plureos GmbH
in Hamburg, Germany